ORGANOSILICON AND BIOORGANOSILICON CHEMISTRY:
Structure, Bonding, Reactivity and Synthetic Application

Editor:

HIDEKI SAKURAI
Professor of Chemistry,
Tohoku University, Sendai, Japan

Published for the
INTERNATIONAL UNION OF PURE AND APPLIED CHEMISTRY

ELLIS HORWOOD LIMITED
Publishers · Chichester

Halsted Press: a division of
JOHN WILEY & SONS
New York · Chichester · Brisbane · Toronto

7227-0585

CHEMISTRY

First published in 1985 by
ELLIS HORWOOD LIMITED
Market Cross House, Cooper Street, Chichester, West Sussex, PO19 1EB, England

The publisher's colophon is reproduced from James Gillison's drawing of the ancient Market Cross, Chichester.

Distributors:

Australia, New Zealand, South-east Asia:
Jacaranda-Wiley Ltd., Jacaranda Press,
JOHN WILEY & SONS INC.,
G.P.O. Box 859, Brisbane, Queensland 40001, Australia

Canada:
JOHN WILEY & SONS CANADA LIMITED
22 Worcester Road, Rexdale, Ontario, Canada.

Europe, Africa:
JOHN WILEY & SONS LIMITED
Baffins Lane, Chichester, West Sussex, England.

North and South America and the rest of the world:
Halsted Press: a division of
JOHN WILEY & SONS
605 Third Avenue, New York, N.Y. 10158, U.S.A.

© 1985 International Union of Pure and Applied Chemistry
IUPAC Secretariat: Bank Court Chambers, 2-3 Pound Way, Cowley Centre, Oxford OX4 3YF, UK

British Library Cataloguing in Publication Data
International Symposium on Organosilicon Chemistry (7th : 1984 : Kyoto)
Organosilicon and bioorganosilicon chemistry: structure, bonding, reactivity and synthetic application : proceedings of the Seventh International Symposium on Organosilicon Chemistry, Kyoto, Japan, 9–14 September 1984.
1. Organosilicon compounds
I. Title   II. Sakurai   III. International Union of Pure and Applied Chemistry, Inorganic Chemistry Division   IV. International Union of Pure and Applied Chemistry, Organic Chemistry Division   V. Nihon Kagakkai   VI. Nihon Gakujutsu Kaigi
547'.08   QD412.S6

LIBRARY OF CONGRESS Card No. 85–926

ISBN 0–85312–845–6 (Ellis Horwood Limited)
ISBN 0–470–20188–6 (Halsted Press)

Printed in Great Britain by The Camelot Press, Southampton.

Table of Contents

FOREWORD

Hideki Sakurai Department of Chemistry, Faculty of Science, Tohoku University
Sendai 980, Japan

The growth in the organosilicon chemistry of the past twenty years is really explosive. The synthesis and characterization of stable solid compounds containing Si=C and Si=Si double bonds made a dream of long standing an actuality. The recent advent of highly reactive organosilicon intermediates stimulated interest of not only experimental but also theoretical chemists. The "boom" in the application of organosilicon compounds in organic synthesis still continues. The developments in the silicon industry are being actively assessed, and it is now evident that organosilicon compounds can find new ways of applications such as in pre-ceramics and in electronic materials.

The Seventh International Symposium on Organosilicon Chemistry was held in Kyoto, Japan in September, 1984 under these circumstances and the Organizing Committee have tried to bring together papers on as many aspects of organosilicon chemistry as possible in this Symposium. As a result, more than 140 papers were presented and more than 500 people participated in the Symposium. This book is the complete compilation of 27 Plenary and Invited Section Lectures. The contributors are drawn both from Universities and from industry in order to cover a wide variety of topics in organosilicon chemistry. I hope that this book will provide a review on the present state of organosilicon chemistry.

PART 1
STABLE SILICON DOUBLE BONDS

CHAPTER 1

THE DISILENES AND THEIR DERIVATIVES

Robert West*, Mark J. Fink, Michael J. Michalczyk, Douglas J. De Young and Josef Michl‡
Department of Chemistry, University of Wisconsin Madison, WI 53706 U.S.A.
‡ Department of Chemistry, University of Utah, Salt Lake City, UT 84112 U.S.A.

INTRODUCTION

The close relationship between silicon and carbon has led to many attempts over the years to generate multiple bonds to silicon. Early in this
century Kipping [1] carried out experiments directed
toward the synthesis of compounds containing silicon-silicon double bonds (disilenes). The first
evidence for the existence of disilenes as transient
intermediates was provided in the pioneering re-
searches of Peddle and Roark [2], who thermolyzed
bridged disilane derivatives and observed that the
tetramethyldisilene fragment was transferred to
trapping agents such as anthracene (Equation 1).
Over the next decade a number of studies were
published in which the results were interpreted in
terms of disilenes as intermediates [3]. The first
stable disilene, tetramesityldisilene (1) was
finally isolated in 1981 [4], and at least seven
additional stable or nearly stable compounds in
this class have been reported within the past three
years.

This rapidly expanding field of disilene chemistry was treated recently in a review [5] which
covered the synthesis, spectroscopy, structure and
chemistry of this new class of substances as reported in the literature through early 1983. Since a

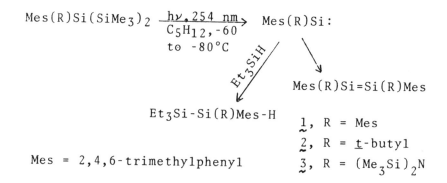

$$Me_2Si{-}SiMe_2$$

$$\longrightarrow \quad + \quad [Me_2Si=SiMe_2]$$

$$\underline{1} \qquad\qquad\qquad\qquad\qquad\qquad C_{14}H_{10}$$

$$Me_2Si{-}SiMe_2 \qquad\qquad\qquad (1)$$

full review of this area seems redundant at this
time, this paper will cover in detail only those
developments occurring after the previous review
[5], with particular emphasis on the work done at
the Universities of Wisconsin and Utah.

SYNTHESIS AND STRUCTURES OF DISILENES

Tetramesityldisilene (1) was initially synthesized by dimerization of dimesitylsilylene, produced by the photolysis of 2,2-dimesitylhexamethyl-
trisilane [4]. As shown in Scheme 1, the same

Scheme 1

$$Mes(R)Si(SiMe_3)_2 \xrightarrow[\substack{C_5H_{12},-60 \\ to\ -80°C}]{h\nu,254\ nm} Mes(R)Si:$$

$$Et_3Si{-}Si(R)Mes{-}H$$

$$Mes(R)Si=Si(R)Mes$$

$$Mes = 2,4,6\text{-trimethylphenyl}$$

$$\underline{1},\ R = Mes$$
$$\underline{2},\ R = \underline{t}\text{-butyl}$$
$$\underline{3},\ R = (Me_3Si)_2N$$

route has been used to obtain disilenes 2 and 3
[6]. The intermediacy of the silylenes Mes(R)Si:
was established by trapping them with triethyl-
silane, or by carrying out the photolysis at 77K in
hydrocarbon glass [7], under which conditions the
silylenes are stable and can be observed directly;
melting of the matrix then produces 1-3.

Photolysis of cyclotrisilanes has also been
found to yield disilenes. Compounds made by this
route include the stable compounds 2, 4 and 5 [8],
and the somewhat less stable tetraalkyldisilenes 6,
7 and 8, which have been obtained only in solution
[9,10,11]. Compound 6 has also been produced by a
retro-Diels-Alder reaction [11] and by chlorine
elimination from t-Bu$_2$SiCl-ClSit-Bu$_2$ [13].

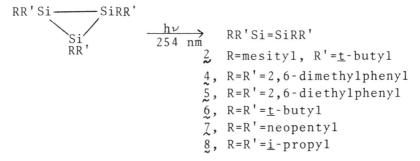

RR'Si————SiRR'

         Si
        / \
      RR'

$\xrightarrow[254\ nm]{h\nu}$ RR'Si=SiRR'

2, R=mesityl, R'=t-butyl
4, R=R'=2,6-dimethylphenyl
5, R=R'=2,6-diethylphenyl
6, R=R'=t-butyl
7, R=R'=neopentyl
8, R=R'=i-propyl

For disilenes to be stable and hence isolable,
it is evidently necessary that the silicon atoms
bear bulky substituents, to protect the Si=Si
double bond against polymerization. This requirement is quite specific; replacement of t-butyl by
i-propyl in 2 gives a disilene which is not isolable. In addition, all of the disilenes isolated
to date contain at least two aromatic rings, although the significance of this observation is not
yet clear.

Structures have been determined for 1, trans-2
and 5 by x-ray crystallography. The Si-Si bond
lengths are near 215 pm for all three compounds, or
about 20 pm shorter than for a normal Si-Si single
bond. However the conformation about the Si=Si
double bond is unique for each of the three compounds. In 1, the silicon atoms are mildly anti-
pyramidalized, with an angle between the C-Si-C
plane and the Si-Si bond of 18°. There is also a
slight (5°) twist at the Si-Si bond [14,15]. In 5,

the arrangement at silicon is planar, but the twist
angle is larger, 10° [8b]. Finally <u>trans</u>-2̰ has a
fully planar, untwisted conformation, as shown in
Figure 1, of the sort found in typical alkenes [15].

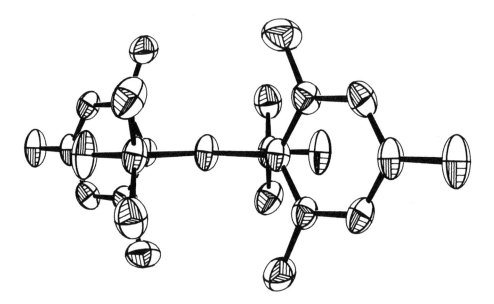

Figure 1 — ORTEP diagram of crystal structure of *trans*-2, looking down the Si-Si bond.

Evidently the Si-Si double bond can easily be de-
formed by twisting or pyramidalization, so that
crystal packing forces may be decisive in determining the conformation in the solid state.

CIS-TRANS ISOMERS

Of the disilenes so far studied, 2̰ and 3̰ can
exist as geometrical isomers [16,17]. Compound 2̰,
produced by photolysis at -80°C in pentane (Equation 1) crystallizes directly as a 98:2 <u>trans</u>:<u>cis</u>
mixture, from which the pure <u>trans</u> isomer can be
obtained by recrystallization as a pale yellow
solid. In the photolysis to produce 3̰, no precipi-
tation takes place, but evaporation of the pentane
solvent and recrystallization gives mainly the
thermodynamically unstable <u>cis</u> isomer (3:97 <u>trans</u>:

cis). When a solution of cis-3 is simply allowed
to stand at 25°C, gradual transformation takes place
to give the equilibrium mixture, 94:6 trans:cis,
from which the pure trans isomer can be obtained.
The two stereoisomers of 3 each form orange-red
crystals.

Pure cis-2 has not yet been isolated, but a
mixture enriched in the cis isomer can be produced
by photolysis. Irradiation of either 2 or 3 at 254
nm produces photostationary mixtures, with trans:
cis ratios 63:37 for 2 and 33:67 for 3. Reversion
of cis-2 to the equilibrium mixture, 98:2 trans:
cis, also takes place in solution at room temperature, but more slowly than for 3.

From a study of the rates of cis-trans interconversion of 2 and 3 as a function of temperature,
the activation energies (ΔE_1) for these processes
have been determined [17]. This value can be taken
as one measure of the strength of the Si-Si π-bond.
For 2, ΔE_1 is 31.3 Kcal mol^{-1}, in good agreement
with molecular orbital calculations which predict a
π-bond energy near 30 Kcal mol^{-1} for disilenes [18].
This is about one-half as large as the value for
typical olefins, which shown ΔE_a near 60 Kcal mol^{-1}.
(However, some highly hindered olefins have ΔE_a
values close to those for 2) [19].

For disilene 3, the ΔE_1 for cis-trans interconversion is significantly lower, 25.4 Kcal mol^{-1}.
The difference between the activation energies of
2 and 3 possibly reflects electron-donation from
nitrogen to the silicon π-system, which could either
weaken the Si-Si π bond or decrease the energy of
the transition state for bond rotation. However,
the structural and conformational differences between 2 and 3 doubtless also play a part. Clearly
more data on rotational isomerism in disilenes
would be useful.

The stereoisomerism of disilenes 2 and 3 allows
an investigation of the stereochemistry of disilene
reactions. The readily-available trans isomer of
2 has been used as substrate for several addition
reactions, illustrated in Scheme 2, which generally
proceed in analogy to the reactions of 1 [20].
Addition of alcohols and water to 2 produces a 1:1
mixture of the two possible disastereoisomers (each

Scheme 2

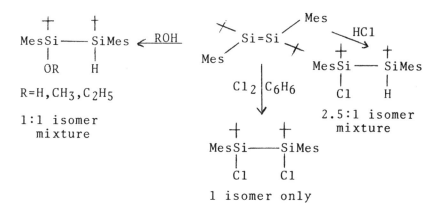

R=H,CH₃,C₂H₅

1:1 isomer
  mixture

2.5:1 isomer
   mixture

1 isomer only

of which consists, of course, of a pair of enantiomers) [21]. These additions therefore very probably proceed stepwise rather than as concerted reactions. Likewise hydrogen chloride adds to 2 to
give a 2:5:1 mixture of diastereoisomers (Scheme 2).
On the other hand addition of chlorine to 2 in benzene yields only one of the two possible diastereoisomers (meso or R,S). This addition reaction must
be concerted, or at any rate must proceed by a
mechanism consistent with retention of stereochemistry. The stereoisomerism of 2 and 3 is also
significant in the oxidation of disilenes, described in the next section.

OXIDATION PRODUCTS OF DISILENES

Solid, powdered disilenes in general react with
atmospheric oxygen. Oxidation is complete within
minutes for 1 and 3, but only after several hours
for 2. The latter compound is sufficiently inert to
oxygen so that the solid can be transferred from one
reaction vessel to another without special precau-
tions. In each case the oxidation product is a
four-membered ring compound, with alternating Si and
O atoms, 9-11. The oxidations appear to be stereospecific, trans-2 producing only trans-10, and cis-
3 exclusively cis-1 [22]. These compounds can be
most simply represented as cyclodisiloxanes, 9a-
11a, but because of their unusual structures the
novel 9b-11b representations must also be considered.

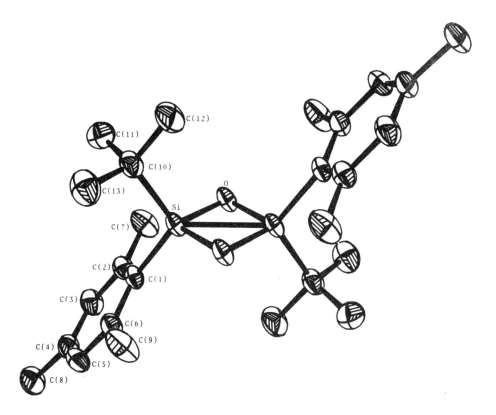

9a, R = Mes   9b

10a, R = _t_-butyl   10b

11a, R = (Me₃Si)₂N   11b

Figure 2 shows an ORTEP diagram for 10. The silicon atoms in the four-membered rings in 9-11

Figure 2 — ORTEP diagram of the structure of 1,3-cyclodisiloxane 10.

are well within bonding distance [23]. The Si-Si
distances in pm are 231 in 9, 235 in 10, and 239 in
11; all of these values are close to the normal Si-
Si single bond distance of 235 pm. The Si-O distances in these molecules are somewhat longer than
normal for a siloxane bond, by about 10 pm. These
facts can be rationalized in terms of the unprecedented structures 9b-11b, in which there is a single
bond between the silicon atoms and electron-defici-
ent Si-O bonds. Thus compounds 9-11 provide a new
problem in chemical bonding theory.

A full understanding of the unique structures
of these molecules will doubtless require additional experiments as well as theoretical calculations. However, the ultraviolet spectra of 9-11
may provide some evidence for direct silicon-silicon bonding, at least in the electronic excited
state. Sakurai has shown that aryldisilanes exhibit
a new band in the uv near 240 nm, not present in
analogous compounds which lack an Si-Si bond. Compounds 9-11 also show a new band in this region,
which decreases in intensity from 9>11>10, parallel-
ing the increase in the Si-Si bond length [23].

$$Mes_2Si \text{——} SiMes_2 \qquad Mes_2Si\text{-}O\text{-}SiMes_2$$

| | | | |
| H | OH | | OH | OH |

12   13

So far we have considered only the oxidation
of 1-3 as solids at room temperature. Reaction of
trans-2 in solution with oxygen takes a different
course, especially at low temperatures. Oxy-
genation of a hexane solution of trans-2 at -78°C
produces initially no 10, yielding instead a mixture
of two new compounds [24]. One is the disiloxirane
14, the first example of this new ring system. The
other product, 15, is an isomer of 10 (Scheme 3).
Epoxide 14 is stable and inert to further oxidation
by O₂, but can be converted to trans-10 by treatment
with m-chloroperbenzoic acid. Compound 15 is
thermally unstable, undergoing quantitative rearrangement to trans-10 at room temperature. At
this time the structure of 15 is not certain; the
1,2-dioxetane formulation 16 seems most likely, but

Scheme 3

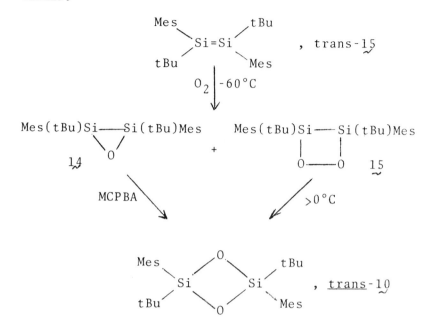

the perepoxide structure 17 cannot be discounted. Both 14 and 15 are formed as single stereoisomers, as indicated by their proton nmr spectra.

```
RR'Si — SiRR'          RR'Si ———— SiRR'
   |      |                    \     /
   O ———— O                      O          R  =  t-butyl
                                 | +  16     R'  =  Mesityl
      15                         O-
```

Reactions of 1 or trans-2 with sulfur have also been carried out, yielding the episulfides 18 and 19 respectively. The reaction of 2 is again stereoselective, producing only one isomer of 19.

$$\text{Mes}\diagdown_{R'}\!\!\diagup^{R}\!\!\text{Si=Si}\diagdown_{\text{Mes}} \xrightarrow{S_8} R(\text{Mes})\text{Si}\!\!-\!\!\text{Si}(\text{Mes})R$$

18, R = Mesityl

19, R = tert-Bu

A related reaction of a disilene leading to three-membered ring formation [25] is the addition of methylene to give the disilacyclopropane 20:

$$R_2Si=SiR_2 \xrightarrow{\ CH_2N_2\ } R_2Si \underset{\diagdown \quad \diagup}{\underline{\hspace{2cm}}} SiR_2$$
$$CH_2$$

20, R = 2,6-dimethylphenyl

SUMMARY

It seems clear that the synthesis of geometric stereoisomers of disilenes has opened an important new area of disilene chemistry. Both the cis-trans interconversions of these isomers and their chemical reactions are beginning to provide significant insight into the nature of silicon-silicon double bonding. Disilene stereoisomers are also proving useful in the investigation of the surprising new oxidation products of the disilenes.

ACKNOWLEDGEMENT   Research was sponsored by the Air Force Office of Scientific Research, Air Force Systems Command, USAF under Contract No. F49620-83-C-0044. An important part of this work was carried out in collaboration with Professor Josef Michl at the University of Utah under National Science Foundation Grant CHE81-811122. The United States Government is authorized to reproduce and distribute reprints for Governmental purposes notwithstanding any copyright notation thereon.

REFERENCES

1. F. S. Kipping, Proc. Roy. Soc., (1911) 27, 143.

2. D. N. Roark and G. J. D. Peddle, J. Am. Chem. Soc. (1971) 94, 5837.

3. For example, T. J. Barton and J. A. Kilgour, J. Am. Chem. Soc. (1974) 96, 7150; ibid. (1976) 98, 7746; ibid. (1976) 98, 7231; W. D. Wulff, W. F. Goure and T. J. Barton, ibid. (1978) 100, 6236; H. Sakurai, T. Kobayashi and Y. Nakadaira, J. Organomet. Chem. (1978) 162, C43; Y.-S. Chen, B. H. Cohen and P. P. Gaspar, ibid. (1980) 195, C1.

4.   R. West, M. J. Fink and J. Michl, Science (1981)
     214, 1343.

5.   R. West, Pure Appl. Chem. (1984) 56, 1, 163.
     For an additional review on stable disilenes
     see R. West, Science, in press.

6.   M. J. Michalczyk, R. West and J. Michl, J. Am.
     Chem. Soc. (1984) 106, 821.

7.   M. J. Fink, M. J. Michalczyk, D. J. De Young,
     K. M. Welsh, R. West and J. Michl, unpublished
     investigations.

8.   a) S. Masamune, Y. Hanzawa, S. Murakami, T.
     Bally and J. F. Blount, J. Am. Chem. Soc.
     (1982) 140, 1150; b) S. Masamune, S. Murakami,
     J. T. Snow, H. Tobita and D. J. Williams,
     Organomet. (1984) 3, 333; c) S. Murakami, S.
     Collins and S. Masamune, Tetrahedron Lett.
     (1984) 2131.

9.   H. Watanabe, T. Okawa, M. Kato and Y. Nagai,
     J. Chem. Soc. Chem. Commun. (1983) 781.

10. A. Schäfer, M. Weidenbruck, K. Peters and H.-G.
     von Schnering, Angew. Chem. (1984) 96, 311.

11. S. Masamune, H. Tobita and S. Murakami, J. Am.
     Chem. Soc. (1983) 105, 6524.

12. S. Masamune, S. Murakami and H. Tobita, Organo-
     met. (1983) 2, 1464.

13. P. Boudjouk, B.-. Han and K. R. Anderson, J.
     Am. Chem. Soc. (1982) 104, 4992.

14. M. J. Fink, M. J. Michalczyk, K. J. Haller,
     R. West and J. Michl, J. Chem. Soc. Chem.
     Commun. (1983) 1010.

15. M. J. Fink, M. J. Michalczyk, K. J. Haller, R.
     West and J. Michl, Organomet. (1984) 3, 793.

16. M. J. Michalczyk, R. West and J. Mich., J. Am.
     Chem. Soc. (1984) 106, 821.

17. M. J. Michalczyk, J. Michl and R. West, un-
     published studies.

18. R. Daudel, R. E. Kari, R. A. Poirier, J. D. Goddard and J. G. Czismadia, J. Mol. Struc. (1978) 50, 115; R. A. Poirier and J. D. Goddard, Chem. Phys. Lett. (1981) 80, 37.

19. H. -O. Kalinowski and H. Kessler in "Topics in Stereochemistry", N. L. Allinger, E. L. Eliel, Eds., Wiley, New York, 1973, Vol. 7, p. 300.

20. M. J. Fink, D. J. De Young, R. West and J. Michl, J. Am. Chem. Soc. (1983) 105, 1070.

21. D. J. De Young and R. West, unpublished studies.

22. M. J. Fink, K. J. Haller, R. West and J. Michl, J. Am. Chem. Soc. (1984) 106, 822.

23. M. J. Michalczyk, M. J. Fink, K. J. Haller, R. West and J. Michl, unpublished studies.

24. M. Michalczyk, R. West and J. Michl, in press.

25. S. Masamune, S. Murakami, M. Tobita and D. J. Williams, J. Am. Chem. Soc. (1983) 105, 7776.

CHAPTER 2

NEW PHOTOCHEMICAL BEHAVIOR OF SILENES DERIVED FROM ACYLSILANES

A. G. Brook Lash Miller Chemical Laboratories, University of Toronto, Toronto, Canada M5S 1A1

For as long as organosilicon compounds have been studied, the complete absence of examples of species containing multiple bonds to silicon (Si=C, Si=O, Si=Si) has provided a challenge to both synthetic and theoretical chemists. Over the years there has been considerable effort expended both in trying to synthesize such compounds, and in developing explanations for their apparent lack of stability. It was not until 1967 that compelling evidence was presented that a silene might exist [1], and in recent years these challenges have been met with the synthesis and characterization of relatively stable solid compounds containing Si=C bonds (silenes or silaethylenes) [2,3,4] and Si=Si bonds (disilenes) [5,6].

Our studies of relatively stable silaethylenes developed from our studies of acylsilanes, R_3SiCOR, a class of compound which showed unusual photochemical properties including a facile rearrangement to siloxycarbenes at wavelengths of light greater than 360 nm [7] (eq. 1) and from our studies of β-ketosilanes, which readily thermally isomerize to silyl enol ethers [8] (eq. 2).

$$R_3Si\overset{\displaystyle O}{\overset{\displaystyle \|}{C}}R' \xrightarrow{h\nu} R_3SiO\ddot{C}R' \qquad (1)$$

$$R_3SiCH_2\overset{\overset{O}{\|}}{C}R' \overset{\Delta}{\longrightarrow} R_3Si\underset{\underset{CH_2}{\|}}{O}CR' \tag{2}$$

Both reactions reflect the well-established tendency for silicon-oxygen bond formation to occur at the expense of weaker silicon-carbon bonds.

Extension of these studies to acyldisilanes, which have the structural features of both acyl- and β-keto-silanes, showed that photolysis of these compounds gave reactions involving attack of oxygen at both the α and β silicon atoms, leading either to disilyloxycarbenes, or to silenes, the latter route being the major pathway under certain conditions [7] (eq. 3).

$$R_3SiSiR_2\overset{\overset{O}{\|}}{C}R' \longrightarrow R_3SiSiR_2\overset{..}{O}CR' + R_2Si=C\overset{OSiR_3}{\underset{R'}{\diagup}}$$

Acylpolysilanes, particularly the tris(trimethylsilyl)acylpolysilanes, were found to give silenes exclusively under very mild conditions (λ > 360 nm, T = 0° or less) which were characterized by their facile addition of alcohols, dienes (Diels Alder and ene products) and other reagents [3,9]. One unusual feature of these silenes was the formation of head-to-head cyclic dimers [9,10] as distinct from the head-to-tail cyclic dimers normally formed by silenes.

In some cases (e.g., R' = Me) acyclic dimers were isolated whose formation was best explained as arising from coupling of two silene molecules at their silicon ends to form a 1,4-diradical which subsequently underwent disporportionation (scheme 1).

Further study of this series of silenes has provided a number of interesting results which vary significantly from compound to compound depending on the nature of R'. Thus for compounds where R' is aliphatic:

(a)　when R' = Et the formation of both the Z and E geometric isomers of the acyclic head-to-head dimers is the only pathway observed: only the acyclic dimer is formed when R' = Me.

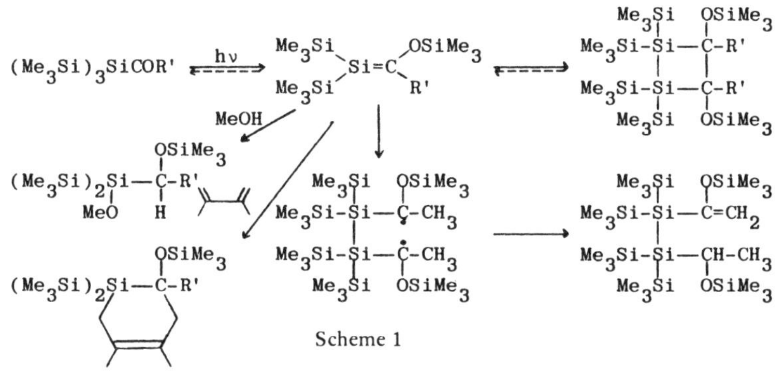

$(Me_3Si)_3SiCOR'$ $\xrightleftharpoons{h\nu}$ $\begin{array}{c} Me_3Si \\ Me_3Si \end{array} Si=C \begin{array}{c} OSiMe_3 \\ R' \end{array}$ $\rightleftharpoons$ $\begin{array}{ccc} Me_3Si & & OSiMe_3 \\ Me_3Si-Si & - & C-R' \\ Me_3Si-Si & - & C-R' \\ Me_3Si & & OSiMe_3 \end{array}$

MeOH

OSiMe$_3$
$(Me_3Si)_2Si-C-R'$
MeO   H

$\begin{array}{cc} Me_3Si & OSiMe_3 \\ Me_3Si-Si-C-CH_3 \\ Me_3Si-Si-C-CH_3 \\ Me_3Si & OSiMe_3 \end{array}$

$\longrightarrow$

$\begin{array}{cc} Me_3Si & OSiMe_3 \\ Me_3Si-Si-C=CH_2 \\ Me_3Si-Si-CH-CH_3 \\ Me_3Si & OSiMe_3 \end{array}$

OSiMe$_3$
$(Me_3Si)_2Si-C-R'$

Scheme 1

(b) when R' = i-Pr, formation of the 1,2-disila-cyclobutane is the major process, but this dimer does not readily dissociate back to monomer, in contrast to the case where R' = t-Bu, where the solid dimer, characterized by x-ray crystal structure, ir, ^{1}H, ^{13}C, ^{29}Si nmr etc. [10] readily dissociates to give an equilibrium mixture with the monomer on dissolving in an inert solvent, undoubtedly because severe steric interaction between the bulky substituents on the two ring-carbon atoms destabilizes the ring. Such severe steric strain evidently is lacking in the isopropyl analog.

(c) when R' = very bulky groups such at CEt$_3$ or 1-adamantyl (C$_{10}$H$_{15}$), dimer formation is completely suppressed, and the monomers have been isolated as solids and characterized by a variety of spectroscopic techniques [3,11]. The crystal structure of the silene (R' = 1-adaman-tyl) fig. 1, clearly shows a molecule with a silicon-carbon double bond length of 1.764 Å, which is found to be twisted about the double bond by about 15°, possibly due to crystal packing forces, or to steric interactions.

Samples of the solid silenes have survived without observable change under argon for over three years, although in the presence of air they immedi-ately react in a puff of smoke to give silanone trimer and silyl carboxylate ester oxidation products.

BOND LENGTHS (Å) AND ANGLES

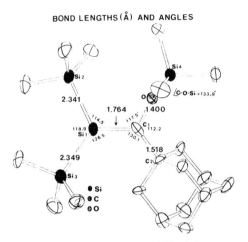

Figure 1 — ORTEP diagram of silaethylene

The remarkable and unprecedented stabilities of the silenes with this substitution pattern can be attributed in part to steric effects, where the bulky substituents inhibit dimerization and reactions with other reagents, and in part to electronic effects where the dominant substituents, Me₃SiO on carbon and Me₃Si on silicon, alter the polarization of the double bond both through inductive effects and through resonance contributions from structures $\underline{B}$ and $\underline{C}$, resulting in a polarization of the double bond much reduced from that to be expected as a result of the differences in electronegativity between silicon and carbon (Si-C). Calculations by Apeloig [12]

confirm the reinforcing influences of a silyl group on silicon and an oxygen on carbon in stabilizing a silicon-carbon double bond kinetically via electronic effects. If the polarization in the π system opposes the normal polarization of the σ system the result can be expected to be a relatively unpolarized double bond not prone to dimerization which is driven by polar effects.

The experimental data are in accord with the above interpretation. The length of the Si=C double bond, 1.764 Å, is significantly shorter than a Si-C single bond, but somewhat longer than the value calculated for a simple silene (1.69-1.72 Å) and hence consistent with some single bond character. The remarkably deshielded position of the sp^2-hybridized carbon, ≈214 ppm below TMS, is well below the normal range of sp^2-hybridized carbon in alkenes, is reminiscent of the carbons in the Fischer transition metal-carbene complexes, and is consistent with some positive charge on carbon. The chemical shift of the sp^2-hybridized silicon, 40-55 ppm below TMS, represents an unusual position for silicon bearing silyl groups but is consistent with a silicon atom which is sp^2-hybridized. However the negative charge attributed to silicon in structures B and C infers that the silicon should be relatively shielded. The absence of appropriate standards has made comparison of chemical shifts difficult until recently, when West [13] published ^{29}Si chemical shifts for sp^2-hybridized silicon atoms in several disilenes. These shifts lie in the range δ 49-95 ppm: hence the sp^2-hybridized silicon atoms in our silenes do appear to be relatively shielded, although part of the difference from West's results must be attributed to the different substituents attached to silicon in the two systems.

We have begun to explore the importance of the roles played by certain structural features associated with our family of silenes, particularly the silyl groups attached to the doubly-bonded silicon, the siloxy group on the doubly-bonded carbon, as well as the extent to which our general route to silicon-carbon double bonds can be extended to other systems.

Becker and coworkers [14] have already established that acylsilylphosphines or arsenes can be converted thermally to doubly-bonded species (eq. 4).

$$Me_3Si \diagdown \atop R \diagup MCOR' \xrightarrow{\Delta} R-M=C \diagup OSiMe_3 \atop \diagdown R' \qquad (4)$$

M = P, As

It seemed possible that a photochemical route similar to that used to form silenes might yield the related germenes, since in the tris(trimethylsilyl)acylgerm-enes the essential β-silyl to carbonyl-oxygen spacial relationship exists (eq. 5). However, photolysis

$$(Me_3Si)_3GeCOC_{10}H_{15} \xrightarrow{h\nu} (Me_3Si)_2Ge=C\begin{smallmatrix} OSiMe_3 \\ \\ C_{10}H_{15} \end{smallmatrix} \quad (5)$$

$$\longrightarrow (Me_3Si)_3Ge\cdot + \cdot COC_{10}H_{15}$$

of the acylgermane failed to yield any products which suggested that a germene had been formed. A multitude of reaction products were formed, some of which seemed to be oligomeric, and the evidence obtained suggested that homolytic cleavage of the acylgermane was the major initial step. This was readily discernable when the photolysis was conducted in carbon tetrachloride, since tris(trimethylsilyl)chlorogerm-ane and adamantoyl chloride were major products. However some of the products underwent secondary photochemical reactions explaining the overall complexity of the reaction system.

As one test of the importance of an oxygen atom, or other donor atom with lone pairs, on the sp^2-hybridized carbon of silenes in affecting their stabilities and reactions, it appeared useful to attempt to photolyze a silanecarboxylate ester. This would lead to a silene with two oxygen-bearing groups on carbon if the photoisomerization occurred (eq. 6).

$$(Me_3Si)_3Si\overset{\overset{\textstyle O}{\|}}{C}OR' \xrightarrow[\Delta]{h\nu\ or} (Me_3Si)_2Si=C\begin{smallmatrix} OSiMe_3 \\ \\ OR' \end{smallmatrix} \quad (6)$$

$$\xrightarrow{\Delta} (Me_3Si)_3SiOR' + CO$$

R' = Me, Me_3Si, Ph_3Si

Unfortunately, while the esters were readily prepared they failed to photoisomerize, probably because the weak n-π* absorption of the ester is hidden under the general polysilyl absorption [15]. Attempts at thermal isomerization to the silene resulted in the alternative well-known thermal decarbonylation reaction [16] so that this approach to bis-oxy-substituted silenes failed and some other approach will be necessary.

The most interesting recent development in our silene research has come from studies where one of the two trimethylsilyl groups on the silicon of the silene has been replaced by a bulky non-silicon group such as t-butyl or phenyl.

Bis(trimethylsilyl)-t-butyladamantoylsilane 1 was readily prepared from t-butyltris(trimethylsilyl)-silane by the usual methyllithium cleavage and acid chloride coupling reactions. Photolysis in methanol appeared to proceed normally giving rise to two isomeric silene-methanol adducts in roughly 3:1 proportions (scheme 2). Photolysis in inert solvents gave rise to at least one silene (δ ^{13}C = 195.6; ^{29}Si = 73.8 ppm) but the existence of a second geometric isomer which could be anticipated, has not yet been absolutely confirmed largely because of other signals which developed during the photolysis (see below). Note that replacement of one silyl group by a t-butyl leads to significant deshielding of the sp^2-hybridized silicon atom when compared to the original adamantylsilene, δ ^{29}Si = 41.4 ppm.

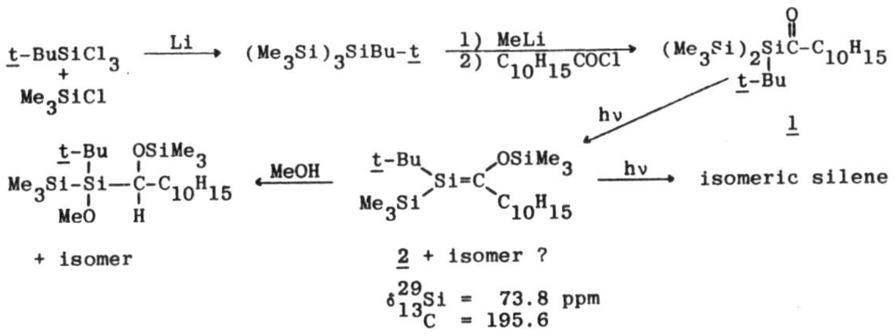

Scheme 2

In the process of photolyzing to complete the isomerization of acylsilane 1 to silene 2 it was observed that new nmr signals developed fairly rapidly which could be attributed to a new silene being generated photochemically at the expense of 2 (these signals did not develop from 2 in the absence of radiation).

This new silene gave methanol adducts different from 3 or 4 and clearly was the result of extensive rearrangement of silene 2. Thus the ^{1}H and ^{13}C nmr data indicated that the three silicon atoms present with ^{29}Si nmr signals at 6.4, 8.6 and 126.5 ppm had respectively two, three and one attached methyl groups. Furthermore, the ^{29}Si nmr data suggested the absence of silicon-silicon bonds in the new silene since these normally result in chemical shifts which lie to high field of TMS [11]. The signal at δ 126.5 ppm is the most deshielded value for sp^2-hybridized silicon reported to date, suggesting the presence of one or more strongly deshielding groups (e.g., oxygen). This new silene was very stable, existing in solution at room temperature for up to two months without change, although it ultimately dimerized to a white precipitate. The crystal structure of this material 6 has been obtained and was found to have a head-to-tail dimeric structure as shown in scheme 3. The structure of 5, the monomer from which 6 evidently is formed, appears consistent with the experimental data obtained for the silene formed on prolonged photolysis of 2.

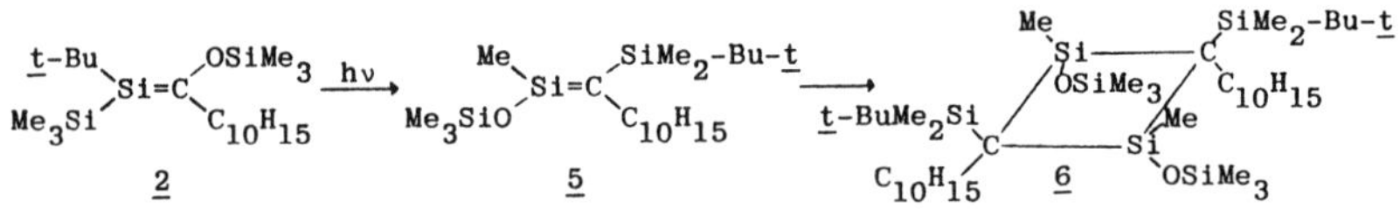

	2	5	6
$\delta^{29}Si$ =	73.8 ppm	126.5 ppm	7.5 ppm
	12.4	8.6	4.3
	-13.1	6.4	2.5
$\delta^{13}C$ =	195.6	118.1 ?	

Scheme 3

That 6 is a head-to-tail dimer of 5 is not surprising since the precursor 5 lacks both the special features of silyl groups on silicon and a siloxy group on carbon that are believed to favour the head-to-head dimerization observed for analogs of 2.

It is obvious that very deep-seated rearrangments have taken place in the isomerization of the silene 2 to the silene formed in solution and presumed to have the structure 5. We are attempting to further characterize 5 both by chemical reactions and by a study of the ^{13}C-^{29}Si spin-spin coupling constants, which may help to confirm the structure of the silene, which if a solid, will be isolated for crystal structure determination.

One possible sequence of reactions which would explain the known results is shown in scheme 4 where the initially formed silene 2 isomerizes via a disilacyclopropane 7 which reopens to the silene 5. This complex sequence of 1,2- shifts is only one way of accounting for the rearrangement of 2 and further studies of the details of the interconversion are underway.

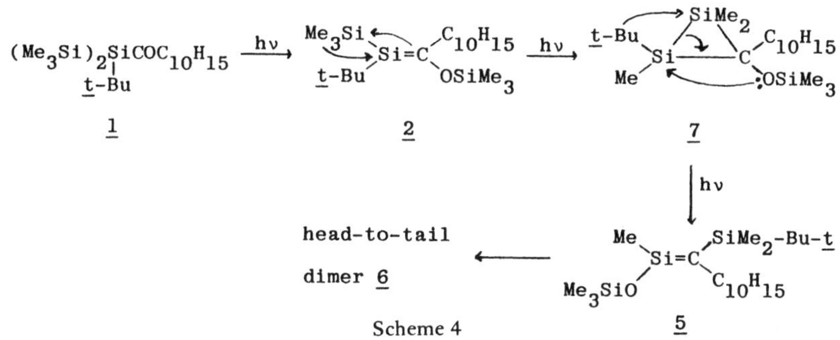

Scheme 4

This extraordinary behavior is not unique since the acylsilane related to 1, but in which t-butyl has been replaced by phenyl, also yields unusual results. While direct photolysis in methanol appears normal, yielding two diastereomeric (?) methanol adducts, photolysis in inert solvents appears to give fairly rapidly two different silenes, judged on the basis of their nmr chemical shifts to be the result of photorearrangement of the initially formed silenes. These latter silenes dimerize quite rapidly to give stable solids believed to be head-to-tail dimers, but which have structures different from 6.

It is very clear that the replacement of a silyl group in the original stable adamantylsilene (which itself survives further extended photolysis without change) with a hydrocarbon group results in strikingly different photochemical behavior. The results bridge, in an extreme way, the unusual head-to-head dimerization behavior observed for many of the silenes derived from acylpolysilanes with the more common head-to-tail dimerization normally observed for silenes lacking silicon on sp^2-hybridized silicon or oxygen on sp^2-hybridized carbon. Further exploration of these remarkable reactions is being carried out.

ACKNOWLEDGEMENT

I wish to acknowledge the splendid work of many associates, and in particular Kim Baines, Kazem Safa, Paul Lickiss and Herbert Söllradl, in carrying out much of the new work described here, and the Natural Science and Engineering Research Council of Canada who provided the financial support for this research.

REFERENCES

1. Guselnikov, L.E.; Flowers, M.C. J. Chem. Soc. Chem. Commun. 1967, 864.
2. Brook, A.G.; Abdesaken, F.; Gutekunst, B.; Gutekunst, G.; Kallury, R.K.M. J. Chem. Soc. Chem. Commun. 1981, 191.
3. Brook, A.G.; Nyburg, S.C.; Abdesaken, F.; Gutekunst, B.; Gutekunst, G.; Kallury, R.K.M.; Poon, Y.C.; Chang, Y-M.; Wong-Ng, W. J. Am. Chem. Soc. 1982, 104, 5667.
4. Wiberg, N.; Wagner, G. Angew. Chem. Int. Ed. Engl. 1983, 22, 1005. .
5. West, R.; Fink, M.J.; Michl, J. Science, 1981, 214, 1343.
6. Masamune, S.; Hanzawa, Y.; Murakami, S.; Bally, T.; Blount, J.F. J. Am. Chem. Soc. 1982, 104, 1150.
7. Duff, J.M.; Brook, A.G. Can. J. Chem. 1973, 51, 2869.
8. Brook, A.G.; MacRae, D.M.; Bassindale, A.R. J. Organometal. Chem. 1975, 86, 181.
9. Brook, A.G.; Harris, J.W.; Lennon, J.; El Sheikh, M. J. Amer. Chem. Soc. 1979, 101, 83.
10. Brook, A.G.; Nyburg, S.C.; Reynolds, W.F.; Poon, Y.C.; Chang, Y-M.; Lee, J-S.; Picard, J-P. J. Amer. Chem. Soc. 1979, 101, 6750.
11. Brook, A.G.; Abdesaken, F.; Gutekunst, G.; Plavac, N. Organometallics, 1982, 1, 994.
12. Apeloig, Y.; Karni, M. J. Amer. Chem. Soc. in press. We are indebted to Dr. Apeloig for providing us with the details of his calculations in advance of publication.
13. West, R. Pure & Applied Chem. 1984, 56, 163.
14. Becker, G.; Gutekunst, G. Z. Anorg. Allg. Chem. 1980, 470, 144, 157.
15. Brook, A.G.; Yau, L. J. Organometal. Chem. in press.
16. Brook, A.G.; Mauris, R.J. J. Amer. Chem. 1957, 79, 971.

CHAPTER 3

RECENT RESULTS IN CHEMISTRY OF UNSATURATED SILICON COMPOUNDS:$>$Si$=$C$<$, $>$Si$=$N$^-$,$>$Si$=$O

N. Wiberg Institut für Anorganische Chemie der Universität München, FRG

Compounds with $p_\pi p_\pi$ bonded silicon ("unsaturated silicon compounds") are highly unstable as compared to those with $p_\pi p_\pi$ bonded carbon ("unsaturated carbon compounds"). During the last few years we have been attempting to devise simple methods to generate unsaturated silicon compounds of type

$$\begin{array}{c} R \\ \diagdown \\ R' \diagup \end{array} Si=C \begin{array}{c} SiR_3' \\ \diagup \\ \diagdown SiR_3' \end{array} \qquad \begin{array}{c} R \\ \diagdown \\ R' \diagup \end{array} Si=N \diagup SiR_3' \qquad \begin{array}{c} R \\ \diagdown \\ R' \diagup \end{array} Si=O$$

$$\text{Silaethenes} \qquad\qquad \text{Silaketimines} \qquad\qquad \text{Silaketones}$$

under mild conditions [1-8]. Such methods are necessary pre-requisites for an extensive and systematic study of *reactivity*, and for the preparation of *stable derivatives* of the said compounds. Our results in context with generation and reactivity of unstable and stable silaethenes, silaketimines, and sila-ketones of the type shown, which, in extract, were presented on the "Seventh International Symposium on Organosilicon Chemistry" will appear in the near future as a review article [8]. There-fore, in the following, only the more important points of the lecture are mentioned and summarized.

PREPARATION

We found salt elimination to be a suitable method for obtaining *silaethenes* $R_2Si=C(SiR_3')_2$ under mild conditions:

$$-\underset{\underset{X}{|}}{\overset{|}{Si}}-\underset{\underset{M}{|}}{\overset{|}{C}}- \quad \underset{-MX}{\overset{\Delta}{\rightleftharpoons}} \quad {}^{\backslash}Si=C{}^{/}$$

(X = electronegative rest, M = electropositive metal). For example, the precursors $Me_2SiX\text{-}CM(SiMe_3)_2$ with X = halogen and M = lithium loose LiF, LiBr, or LiCl according to:

$$Me\text{-}\underset{\underset{X}{|}}{\overset{\overset{Me}{|}}{Si}}\text{-}\underset{\underset{Li}{|}}{\overset{\overset{SiMe_3}{|}}{C}}\text{-}SiMe_3 \quad \underset{-\ LiX}{\overset{\Delta}{\rightleftharpoons}} \quad \overset{Me}{\underset{Me}{}}Si=C\overset{SiMe_3}{\underset{SiMe_3}{}}$$

at almost $0^{\circ}C$, $-80^{\circ}C$, and $-100^{\circ}C$, respectively, leading to dimethyl-bis(trimethylsilyl)silaethene.

The unsaturated silicon compounds, arising from an elimination of MX, dimerize immediately (as a rule). According to our investigations, bromo substituted starting materials ${}^{\backslash}SiX\text{-}CBr{}^{/}$ are especially useful for the synthesis of the silaethene generators, because they react with lithium organyls even at $-100^{\circ}C$, in a Br/Li-exchange.

If the silaethene $Me_2Si=C(SiMe_3)_2$ is generated from one of the above mentioned "sources" in the presence of $Ph_2C=N(SiMe_3)$, yellow [2+4] and colourless [2+2] cycloadducts are formed, which, in turn, may act above $60^{\circ}C$ as easy to handle "stores" for $Me_2Si=C(SiMe_3)_2$:

The intermediate formation of the silaethene by thermal [2+4] and [2+2] cycloreversion of the adducts can be directly "shown", if the adducts are sublimed at low pressure through a hot quartz tube into the ion source of a mass spectrometer (appearance of the molecule ion peak).

Silatriazolines and silatetrazolines of the following type act as good generators for *silaketimines* $R_2Si=N-SiR_3'$ (thermal [2+3] cycloreversion):

$$(Me_3Si)_2C \underset{N=N}{\overset{\overset{Me_2}{Si}}{\diagdown}} N-R \quad \xrightarrow[- (Me_3Si)_2C=N=N]{ca. -10^\circ C} \quad Me_2Si=N-R$$

$$R = (CMe_3), Ph, SiMe_3, SiMe_2{}^tBu, SiMe^tBu_2, SiPh_3$$

$$R-N \underset{N=N}{\overset{\overset{Me_2}{Si}}{\diagdown}} N-R \quad \underset{\mp RN=N=N}{\overset{ca. 120^\circ C}{\rightleftarrows}} \quad Me_2Si=N-R$$

$$R = (SiMe_3), SiMe_2{}^tBu, SiMe^tBu_2, Si^tBu_3, SiPh_3$$

The former are easily prepared from silaethenes, generated by salt elimination method, and azides RN_3, the latter from sila-ketimines, generated by [2+3] cycloreversion of silatriazolines, and azides RN_3. In an analogous manner, *silaketones*, $R_2Si=0$, can be generated by [2+3] cycloreversion of silaoxatriazolines (prepared from silaketimines and N_2O):

$$R-N \underset{N=N}{\overset{\overset{Me_2}{Si}}{\diagdown}} 0 \quad \xrightarrow[- RN=N=N]{< 80^\circ C} \quad Me_2Si=0$$

$$R = Si^tBu_3$$

REACTIVITY

Unsaturated silicon compounds $R_2Si=Y(SiR_3')_n$ (Y = C, N, O), generated from appropriate starting materials in the presence of reactants ("trapping agents") often combine with these reactants more rapidly, to form secondary products, and then dimerize (or oligomerize, or polymerize). This often provides an opportunity to study the reactivity of these compounds. According to our investigations, the unsaturated systems $R_2Si=Y(SiR'_3)_n$ react with polar a–b single bonds of many compounds forming *insertion products*, combine with a=b–c–H double bond systems to give *ene reaction products*, and add to unsaturated compounds of type a=b, a–b≡c, and a=b–c=d to give [2+2], [2+3] and [2+4] cycloadducts:

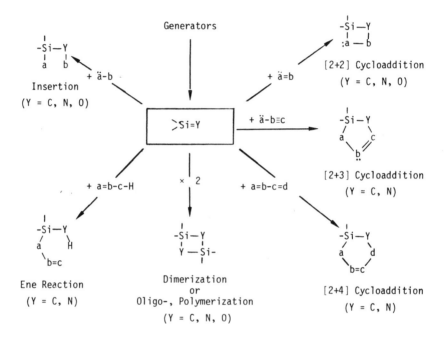

For example, silaethene $Me_2Si=C(SiMe_3)_2$, which, as we were able to demonstrate, is fluxional:

inserts into O–H, N–H, Si–Hal, Si–O, Si–N bonds of alcohols amines, silyl halides, ethers, or amines, respectively; ene reactions are observed, among others, with alkenes or acetone, [2+4] cycloadditions with dienes, [2+3] cycloadditions with azides or dinitrogen oxide, [2+2] cycloadditions with ketones, ketimines, azo and nitrose compounds (for details cf. ref. [8]).

A STABLE SILAETHENE

Detailed kinetic investigations have shown that the simple silaethene $Me_2Si=C(SiMe_3)_2$ is only formed as a short-lived reaction intermediate. With the increasing bulk of the silicon-bonded alkyl groups, the metastability of the unsaturated silicon compound   increased. Thus, we could isolate the silylethene $Me_2Si=C(SiMe_3)(SiMe^tBu_2)$, derived from the above mentioned silaethene by formal substitution of two methyl by tert.butyl groups. These findings parallel those of Brook et al. for silaenolethers $(Me_3Si)_2Si=CR(OSiMe_3)$, which were found to become increasingly stable under normal conditions as R is of sufficient bulkiness (e.g. R = adamantyl) [9].

The moderately stable silaethene $Me_2Si=C(SiMe_3)(SiMe^tBu_2)$, prepared by salt elimination method, with the exception of hindered self-decomposition, shows the same reactivity as un-- stable silaethenes. No single crystals, suitable for X-ray crystallography, of this silaethene has been obtained so far. However, the molecule does form a well crystallizing tetrahydrofuran adduct, the structure of which is as follows:

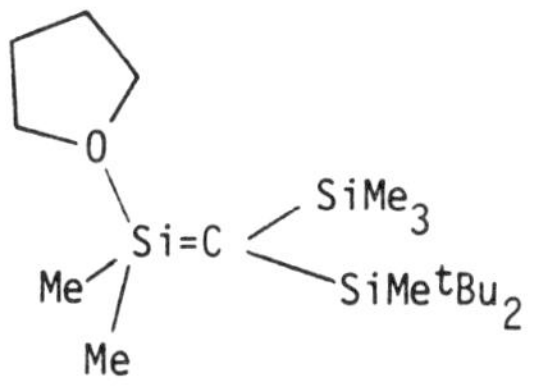

Obviously, it is the unsaturated silicon atom, which is approached by the oxygen atom of the thf molecule, the SiO-distance being comparatively long (1.878 Å). The bond between the central silicon and carbon is drastically shorter (1.747 Å) than SiC single bonds (1.87 - 1.97 Å). The central C-atom is planar and the central silicon atom lies at the apex of a pyramid, if thf is not considered. The non planarity of the unsaturated silicon atom may be inherent of silaethene systems or may be a consequence of thf coordination. With the structural determination of the said thf adduct, the Lewis acidity of unsaturated silicon compounds could be demonstrated for the first time in a direct way. (Beyond that, in the meantime, we were able to isolate a stable tetrahydrofuranat of the silaketimine $Me_2Si=N-Si^tBu_3$.)

The tetrahydrofuranat of $Me_2Si=C(SiMe_3)(SiMe^tBu_2)$ may be sublimed in high vacuum at room temperature with partial decomposition. According to mass spectroscopic results, the gas phase consists of discrete silaethene and thf molecules. Clearly the adduct bonding is not very strong and is reversible. Reactions of the adduct with methanol, acetone, oxygen, azides, dienes yield, apparently by way of adduct dissoziation, the same insertion, ene reaction, cycloaddition products as the analogous reactions with uncomplexed $Me_2Si=C(SiMe_3)(SiMe^tBu_2)$.

1. N. Wiberg and G. Preiner, Angew. Chem. Int. Ed. Engl. 16, 328 (1977).
2. N. Wiberg and G. Preiner, Angew. Chem. Int. Ed. Engl. 17, 362 (1978).
3. N. Wiberg, G. Preiner, and O. Schieda, Chem. Ber. 114, 2087 (1981).
4. N. Wiberg, G. Preiner, O. Schieda, and G. Fischer, Chem. Ber. 114, 3505 (1981).
5. N. Wiberg, G. Preiner, and O. Schieda, Chem. Ber. 114, 3518 (1981).
6. N. Wiberg and G. Wagner, Angew. Chem. Int. Ed. Engl. 22, 1005 (1983).
7. N. Wiberg, G. Wagner, G. Müller, and J. Riede, J. Organomet. Chem., September 1984 ("Kumada issue").
8. N. Wiberg, J. Organomet. Chem., end of 1984.
9. A. G. Brook, S. C. Nyburg, F. Abdesaken, B. and G. Gutekunst, R. Krishna, M. R. Kallury, Y. C. Poon, Y. M. Chang, W. Wong-Ng, J. Am. Chem. Soc. 104, 5667 (1982).

PART 2

**THEORETICAL APPROACH AND
REACTIVE INTERMEDIATES**

CHAPTER 4

MOLECULAR ORBITAL STUDIES OF STRUCTURE AND REACTIVITY OF ORGANOSILICON COMPOUNDS

Keiji Morokuma Institute for Molecular Science, Myodaiji, Okazaki 444, Japan

INTRODUCTION

Last several years many ab initio molecular orbital studies on the structure, stability, spectroscopy and reactivity of organosilicon compounds have been published [1-2]. Such profound activities are based on some recent advances of theoretical methods. The most important factor contributing to this is the development of the energy gradient method. In this method, the gradient of total energy with respect to the nuclear coordinates, gradE, is calculated analytically, at the same time the wavefunction and its energy E is calculated. With the energy gradient in hand, one can easily optimize the geometry of stable polyatomic molecules and also the geometry of the transition state, the saddle point on the multidimensional potential energy hypersurface. The energy gradient method virtually revolutionized the way quantum chemists explore potential surfaces controlling molecular structure and chemical reactions. Optimization of all the geometrical parameters with the Hartree-Fock method for molecules having up to 20 atoms now often falls within a range of practicality, though the computer time required depends very much on the size of basis set adopted. Another contributing factor is advances in methods of calculating electron correlation, and some of them have been made available for application to molecules of reasonable size.

In this lecture I present three applications of
molecular orbital theory to problems of structure
and reactivity of organosilicon compounds. They are
(1) the problem of molecular structure in connection
with the conjugation between Si-Si σ bonds and C≡C
π bonds in tetrasilacyclooctadiyne and
trisilacycloheptadiyne, (2) the barrier for
dimerization in substituted silaethylenes and a
tactic to prevent dimerization by electronic control, and (3) revisit of an old problem of non-least
motion vs. least motion, in comparing dimerization
paths of $CH_2 + CH_2$, $SiH_2 + SiH_2$ and $CH_2 + SiH_2$.

CONJUGATION BETWEEN Si-Si σ BONDS AND C≡C π BONDS IN 3,4,7,8-TETRASILACYCLOOCTA-1,5-DIYNE AND 3,6,7-TRISILACYCLOHEPTA-1,4-DIYNE

Sakurai, Nakadaira, Hosomi, Eriyama and Kabuto
[3] have recently prepared 3,3,4,4,7,7,8,8-
octamethyl-3,4,7,8-tetrasilacycloocta-1,5-diyne,
1a. Based on a qualitative molecular orbital dia-

$R_2Si-C≡C-SiR_2$
$R_2Si-C≡C-SiR_2$

1a : $R=CH_3$
1b : $R=H$

$R_2Si-C≡C-SiR_2$ (to SiR_2 ... $C≡C$)

2a : $R=CH_3$
2b : $R=H$

gram, they proposed strong mixing between Si-Si σ
orbitals and C≡C π orbitals with through-conjuga-
tion. They have shown an enhanced bathochromic
shift of UV spectra as an evidence for this extensive σ-π mixing. They have also prepared
3,3,6,6,7,7-hexamethyl-3,6,7-trisilacyclohepta-1,4-
diyne, 2a, a smallest diyne, in which the C≡C
triple bond is shown to be very much distorted as
judged by ^{13}C NMR.
    In order to shed some light into this unique
series of compounds, we have carried out geometry
optimization and analyzed orbital energies, total
energies and electron distribution for 1b and 2b,
where R=H instead of $R=CH_3$ in 1a and 2a. The fully
optimized geometries with the Hartree-Fock(HF) method and the 3-21G basis set are shown in Fig. 1,
together with results for disilane and acetylene.
The agreement between the calculation for 1b and

the experiment for 1a is excellent. A very small
distortion of both acetylene and disilane units is
observed. The SiSi distance in 1a (2.42₄A calculated, 2.36₀A observed) is 0.03-0.04A longer than
the corresponding distance in SiH₃-SiH₃ (2.38₃A
calculated, 2.33₁A observed by electron diffraction), indicating a slight weakening of the bond.
The optimized geometry for 2b constitutes a prediction. The comparison between 1b and 2b suggests
that 2b is very much distorted, and the strain is
localized mainly at acetylene and isolated silylene
units. In the optimization we have assumed the
highest possible symmetry (D₂ₕ for 1b and C₂ᵥ for
2b). The force constant matrix for out-of-plane
vibrations, obtained by numerical differentiation of
the gradient, at the optimized geometry of 1b gave
all real vibrational frequencies. This confirms
that the molecule has a plane of symmetry containing
all the carbon and silicon atoms.
     Energies, correlation diagram and schematic
representation of the several highest occupied or-

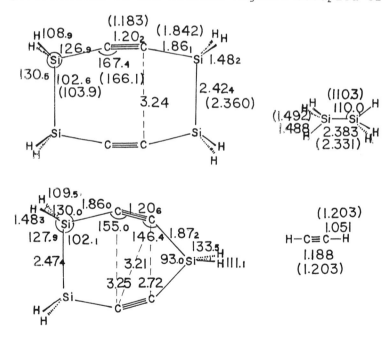

Figure 1 — Optimized geometries of **1b**, **2b**, disilane and acetylene in units of angstroms and degrees. Numbers in parentheses are experimental values for **1a**.

bitals for 1b at the 3-21G HF level are shown in
Fig. 2. One sees clearly that the HOMO (MO44, a_g)
is an antibonding and nearly equal mix of a C≡C π
orbital and an Si-Si σ orbital. One can find a
corresponding bonding mix in MO39 (a_g). The
splitting between the two levels is 3.35eV. If one
assumes a Hückel type interaction among two C≡C π
orbitals and two Si-Si σ orbitals and further
assumes that their orbital energies are the same,
one effectively obtains a cyclobutadiene-like
orbital scheme, where the energy difference between
the highest orbital and the lowest orbital is 4β.
If one interpretes this 4β as corresponding to
above mentioned 3.35eV, the effective resonance
integral between a C≡C π orbital and an Si-Si σ
orbital becomes -0.84eV. Since the orbital energy
of an Si-Si σ bond is not exactly equal to that of a
C-C π bond, this value of effective resonance inte-

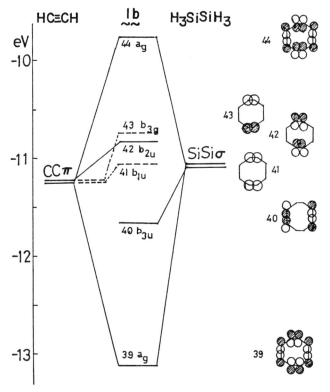

Figure 2 — Energies, correlation diagram, and schematic representation of several highest
occupied orbitals of **1b**.

gral is probably overestimated, but still represents a substantial interaction between them.

Energies, correlation diagram and schematic representation of several highest occupied orbitals for 2b are shown in Fig. 3. One can see similar mixing of Si-Si σ and C≡C π orbitals in the HOMO (MO36, a_g) and MO32 (a_g), though the mixing pattern is more complicated than in 1a due to distortions in constituent units. With the Hückel analogy mentioned above, this system is analogous to allyl and the orbital energy difference is $2\sqrt{2}\beta=-2.40$eV, resulting in an effective resonance integral of −0.78eV.

CONTROLLING BARRIER HEIGHTS FOR DIMERIZATION OF SILAETHYLENES

Silaethylenes are very reactive species and hard to isolate. They dimerize easily to give 1,3-disilacyclobutanes, and their double bonds readily add polar reagents [4]. Our ab initio calculations have shown that the dimerization takes place via a 2_s+2_s cycloaddition path with a very

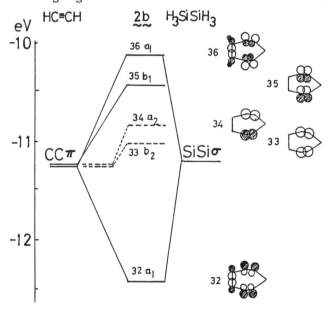

Figure 3 — Energies, correlation diagram, and schematic representation of several highest occupied orbitals of 2b.

low barrier [1]. The breaking of symmetry of π and
π^* orbitals is large enough to invalidate the symmetry rule against $2_s + 2_s$ addition. Silaethylenes
that have been isolated experimentally so far have
rather bulky substituents which should sterically
keep Si=C double bonds away from each other and
thus prevent dimerization [5].

In order to find a guiding principle to control
the barrier height for dimerization electronically
rather than sterically, we have carried out ab
initio calculations for a series of monosubstituted
silaethylenes, SiHX=CHY. Since we are concerned
only with qualitative characteristics of change in
barrier height upon substitution, calculations were
carried out at the HF level with the 3-21G basis
set. The optimized geometry of the transition state
for dimerization of silaethylene (X=H, Y=H) is
shown in Fig. 4, together with that of silaethylene
itself. Assuming that these monomer and transition
state geometries do not change upon introductin of
substituents, we have calculated the barrier height
for dimerization of monosubstituted monomers as well
as monomers' HOMO (π) and LUMO (π^*) energies, frontier densities and total net charges (from gross
atomic population) on Si and C atoms. In the calculation the geometry of substitutent X was taken from
optimized geometry for CH_3X, and the Si-X distance
was assumed to be 0.375A longer than the C-X
distance.

The results for X or Y=H, CH_3, NH_2, OH, F, SH,
CN, CHO, SiH_3 and NO_2 are shown in Table I, together with their inductive and resonance substituent
constants, σ_I and σ_R. One finds among monosubstituted silaethylenes that the barrier is the highest
for Y=NH_2 (31 kcal/mol), followed by Y=OH (22 kcal/
mol), Y=F (16 kcal/mol) and X=SiH_3 (12 kcal/mol).
In SiH_2CHNH_2, however, the optimized monomer geometry has a very long C-Si distance (1.90$_6$A), short
C-N distance (1.31$_4$A), and a non-planar Si moiety,
indicating that this molecule any more does not
have a C=Si double bond and cannot be considered a
substituted silaethylene. One notes on the other
hand that for Y=OH, the optimized monomer geometry
is normal (R(CSi)=1.74$_6$A with planar Si moiety).
There is a good correlation between the calculated
barrier height and the σ_R value of ligand Y on the
carbon atom; a strongly π donating resonance substituent on the carbon atom increases the barrier
height. Not so much profound, there is also a
correlation between σ_I value of ligand X and the

Table I — Unoptimized barrier height ΔE for dimerization, monomers HOMO and LUMO orbital energies and population, total net charge and substituent constants (σ_R and σ_I) for monosubstituted silaethylenes CHX=SiHY.

Y	σ_R	ΔE^a (kcal/mol)	HOMO ε (hartree)	LUMO ε (hartree)	HOMO q Si	HOMO q C	LUMO q Si	LUMO q C	Net Charge Si	Net Charge C
NH$_2$	-0.78	31.4	-.2427	.1284	.583	.305	.392	.642	+0.576	-0.367
OH	-0.68	22.1	-.2923	.0960	.515	.408	.480	.533	+0.653	-0.371
F	-0.59	15.5	-.3175	.0809	.463	.502	.544	.465	+0.757	-0.341
SH	-0.16	8.6	-.2956	.0764	.388	.396	.554	.455	+0.852	-0.997
CH$_3$	-0.07	13.0	-.2967	.0976	.465	.537	.541	.458	+0.788	-0.735
H	0	8.5	-.3125	.0918	.437	.563	.575	.425	+0.808	-0.943
CN	0	0.5	-.3378	.0383	.342	.527	.639	.252	+0.964	-0.750
NO$_2$	0	-3.4	-.3676	.0130	.317	.629	.610	.133	+0.993	-0.554
CHO	0.03	1.2	-.3354	.0421	.366	.535	.614	.200	+0.906	-0.872
SiH$_3$	0.07	4.4	-.3186	.0743	.395	.540	.633	.287	+0.871	-1.195

X	σ_I	ΔE (kcal/mol)	HOMO ε (hartree)	LUMO ε (hartree)	HOMO q Si	HOMO q C	LUMO q Si	LUMO q C	Net Charge Si	Net Charge C
SiH$_3$	-0.11	12.2	-.3139	.0766	.452	.540	.344	.415	+0.602	-0.923
CH$_3$	-0.06	4.7	-.3003	.1114	.422	.589	.635	.397	+1.060	-0.975
H	0	8.5	-.3125	.0918	.437	.563	.575	.425	+0.808	-0.943
NH$_2$	0.11	-3.2	-.2788	.1359	.280	.642	.720	.323	+1.211	-1.051
OH	0.28	-3.1	-.3101	.1079	.337	.629	.660	.355	+1.143	-1.040
SH	0.28	5.5	-.3166	.0878	.345	.548	.653	.395	+0.958	-0.967
CHO	0.35	8.2	-.3246	.0543	.430	.555	.292	.354	+1.032	-0.959
F	0.56	-4.8	-.3256	.0919	.380	.608	.627	.382	+1.167	-1.018
CN	0.61	7.6	-.3330	.0578	.433	.550	.415	.402	+1.158	-0.949
NO$_2$	0.70	5.3	-.3507	.0333	.451	.543	.264	.321	+1.106	-0.948

a Because the geometries are not optimized, in some cases we obtain negative ΔE.

barrier height; a σ donating inductive substituent
on the silicon atom increases the barrier height.
In order to test the effect of double substitution,
a calculation was carried out for a system with Y=F
and X=SiH$_3$; the barrier height is found to be 21.1
kcal/mol and 19.9 kcal/mol when X and Y are cis and
trans, respectively. The increase of barrier for
double substitution 21.1-8.5(unsubstituted)=12.6
kcal/mol, or 19.9-8.5=11.4 kcal/mol is only
slightly more than the sum of barrier changes for
single substitution, i.e., 15.5-8.5=7.0 kcal/mol
for Y=F and 12.2-8.5=3.7 kcal/mol for X=SiH$_3$. The
effect of double substitution is, therefore, more
or less additive. Though we have not actually
determined the transition state for dimerization of
substituted compounds, nor we have carried out high
accuracy calculations (large basis set and electron
correlation) to assess the absolute value of barrier
height, we may propose that silaethylenes stable to
dimerization may be isolated by putting on the
silicon atom one or preferably two π donating substituents such as OR or halogen atoms. Though less
effective, σ donating substituents on the carbon
atom may also help. Though the origin of the
barrier change is not clear, resutls in Table I
suggest that these substituents have made the net
charges on both Si and C smaller as well as they
concentrated the HOMO frontier density more on Si
and the LUMO frontier density more on C.

PATHS OF DIMERIZATION OF CH$_2$ AND SiH$_2$

Dimerization of two singlet methylenes to form

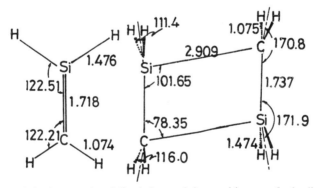

Figure 4 — Optimized geometries of silaethylene and the transition state for its dimerization, in units of angstroms and degrees.

the ground state ethylene had been considered as a
textbook example of non-least motion reaction
path [6]. Recently it has been found, however, from
ab initio calculations that two ground state triplet
methylenes dimerize in the least motion path without
a barrier, whereas two singlet methylenes dimerize
in the least motion path to give a Rydberg excited
state of ethylene [7]. Since silylene has the
singlet ground state as opposed to the triplet
ground state for methylene, its dimerization char-
acteries is of considerable theoretical interest.
Very recently Ohta, Davidson and Morokuma [8] have
studied least motion and non-least motion reaction
paths for CH$_2$+CH$_2$→CH$_2$=CH$_2$, SiH$_2$+SiH$_2$→SiH$_2$=SiH$_2$ and
CH$_2$+SiH$_2$→CH$_2$≡SiH$_2$, starting from both singlet and
triplet state fragments. Here we present a summary
of their results for SiH$_2$ reactions.

Fig. 5 shows potential curves along the D$_{2h}$
least motion path (Curves A, B and C) and along a C$_s$
non-least motion path (Curve D) for dimerization
of SiH$_2$ as functions of SiSi bond length, calculated
with a 4-electron/4-orbital CAS(complete active
space) MCSCF method with a double zeta plus polarization basis set. Curve A, where the HSiH angle and
the SiH distance were fixed at the calculated values
for the ^{1}A$_1$ ground state of SiH$_2$, 93.9° and 1.497A,
respectively, represents the dimerization of singlet
silylenes. Analysis of the wavefunction indicates
that at the first half of reaction the major configuration is, as expected, singlet×singlet, which
increases the energy as the SiSi distance decreases.
Around 3.0A, where a barrier is found, the triplet×
triplet configuration takes over, leading to the
ground state of disilene. In Curve B and C, the
HSiH angle and the SiH distance were fixed at the
calculated values for the ^{3}B$_1$ state of SiH$_2$, 118.0°
and 1.460A, respectively. Since even at this geometry the ^{1}A$_1$ state is lower in energy than ^{3}B$_1$,
Curve B, as in Curve A, at a long distance consists
mainly of singlet×singlet, and becomes triplet×
triplet inside the barrier due to avoided crossing
at around 4.0A. Curve C is the least motion
dimerization path for ^{3}B$_1$, which is a triplet×
triplet at a long distance, goes through an avoided
crossing with Curve B and is adiabatically led to an
excited state of disilene.

The non-least motion path for dimerization of
singlet SiH$_2$ was determined by the HF optimization
as a function of SiSi distance. The potential curve
D along this path shows formation of the ground

state disilene without a barrier. In conclusion,
the ground state singlet SiH_2 dimerizes to give the
ground state disilene with no barrier on a non-
least motion path and with a barrier on the least motion path. The excited state triplet SiH_2 is adiabatically led to an excited state of disilene, but
actually is likely to form the ground state disilene
with no barrier through nonadiabatic transition in
the least motion path.

Fig. 6 sumarizes potential energy curves for
$CH_2+SiH_2 \rightarrow CH_2=SiH_2$. Curve A is the least motion path
for approach of singlet CH_2 and singlet SiH_2 in
their respectively optimum geometries. It goes up
in energy as they come closer until a barrier is
reached where an attractive triplet×triplet configuration takes over to reach the ground state product. Curve B represents the least motion path for
triplet CH_2 + triplet SiH_2. It is higher in energy
than Curve A at infinite separation, but potential
curve is attractive to reach the ground state product with no barrier. Curve D is the non-least motion path for singlet + singlet, which without barrier leads to the ground state product. An interesting observation is that in the non-least motion
path the SiH_2 plane is nearly perpendicular to the
line of approach while the CH_2 plane is parallel to

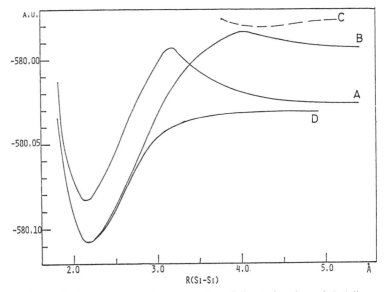

Figure 5 — Potential curves for dimerization of SiH_2 as functions of Si-Si distance.

the line, as if SiH$_2$ is acting as an electron acceptor and CH$_2$ as an electron donor. An energy analysis reveals that in an early stage of reaction the electrostatic interaction is dominant and it is most favorable when a large CH$_2$ dipole is aligned parallel to the line of approach.

CONCLUDING REMARKS

Thanks to development of new methods and availability of faster computers, ab initio molecular orbital methods have become a useful tool for chemistry. The reliability of calculated results improved and the predictability increased. In the coming years the methods are expected to take a significant role in organosilicon chemistry as well as other fields of chemistry. Molecular orbital calculations may help one to conceive a non-existing molecule and examine its structure, stability, reactivity and other properties theoretically before one

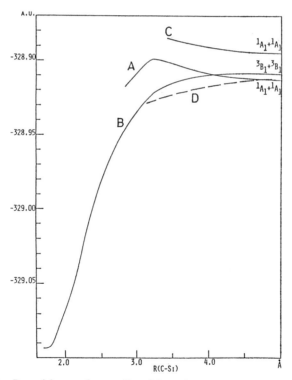

Figure 6 — Potential curves for coupling of CH$_2$+SiH$_2$ as functions of Si-C distance.

actually goes out to laboratory to synthesize. One
may also be able to design or control a chemical
reaction leading to a specific product by calculating in advance transition states and barrier
heights for various proposed reactions. Here col-
laborations between theorists and experimentalists
will be essential.

REFERENCES

1. Hanamura, M.; Nagase, S.; Morokuma, K. Tetra-
   hedron Lett., 1981, 22, 1813; Morokuma, K.; Kato,
   S.; Kitaura, K.; Obara, S.; Ohta, K.; Hanamura,
   M. in "New Horizons of Quantum Chemistry" (
   Löwdin, P.O.; Pullman, B., Eds.), Reidel, 1983,
   221.
2. To mention a few, Hopkinson, A.C.; Lien, M.H.;
   Csizmadia, I.G. Chem. Phys. Lett., 1983, 95,
   232; Hoffmann, M.R.; Yoshioka, Y.; Schaefer, H.F.
   J. Am. Chem. Soc., 1983, 105, 1084; Gordon, M.S.;
   Boudjouk, P.; Anwari, F. J. Am. Chem. Soc.,
   1983, 105, 4972; Gordon, M.S. Theo. Chim. Acta,
   1983, 62. 563; Hopkinson, A.C.; Lien, M.H.
   Theochem, 1983, 9, 153; Nagase, S.; Kudo, T.
   Theochem, 1983, 12, 35; Gordon, M.S.; Gano, D.R.;
   Boatz, J.A. J. Am. Chem. Soc., 1983, 105, 5771;
   Dakkouri, M.; Obserhammer, H. J. Mol. Str.,
   1983, 102, 315; Chandrasekhar, J.; Schleyer,
   P.R.; Baumgartner, R.O.W.; Reetz, M.T. J. Org.
   Chem., 1983, 48, 3453.
3. Sakurai, H.; Nakadaira, Y.; Hosomi, A.; Eriyama,
   Y.; Kabuto, C. J. Am. Chem. Soc., 1983, 105,
   3359.
4. Coleman, B.; Jones, M. Rev. Chem. Intermed.,
   1981, 4, 297; Bertrand, G.; Trinquir, G.;
   Mazerolles, P. J. Organomet. Chem. Library,
   1981, 12, 1.
5. Brook, A.G.; Abdesaken, F.; Gutekunst, B.;
   Gutekunst, G.; Kallury, R.K.M.R. J. Chem. Soc.
   Chem. Commun., 1981, 191; Brook, A.G.; Kallury,
   R.K.M.R.; Poon, Y.C. Organometallics, 1982, 1,
   987.
6. Hoffmann, R.; Gleiter, R.; Mallory, F.B. J. Am.
   Chem. Soc., 1970, 92, 1460.
7. Ruedenberg, K.; Schmidt, M.W.; Gilbert, M.M.;
   Elbert, S.T. Chem. Phys., 1982, 71, 41; Feller,
   D.; Davidson, E.R. J. Phys. Chem., 1983, 87,
   2721.
8. Ohta, K.; Davidson, E.R.; Morokuma, K. to be
   submitted.

CHAPTER 5

RECENT INVESTIGATIONS ON SHORT-LIVED ORGANOSILICON MOLECULES AND MOLECULAR IONS[1]

H. Bock, B. Solouki, P. Rosmus, R. Dammel, P. Hänel, B. Hierholzer,
U. Lechner-Knoblauch and H. P. Wolf, Institute of Inorganic Chemistry,
University of Frankfurt, Niederurseler Hang, D–6000 Frankfurt (M) 50, West Germany

INTRODUCTION: MOLECULAR STATE 'FINGERPRINTS' IN THE INVESTIGATION OF ORGANOSILICON COMPOUNDS

The properties of a molecule depend on its energy and will change especially on acquisition or loss of electrons. The individual state of a molecule or a molecular ion is characterized by energy difference from a preceding or to a subsequent state as well as by the respective charge distribution. This molecular state approach has been and is used by the chemist for his own benefit in many ways, for example, supported by symmetry considerations and more rigorous quantum chemistry calculations to compare 'equivalent states' of 'chemically related', preferably iso(valence)electronic molecules [2] or guided by orbital models to disentangle the multitude of known compounds and to rationalize their numerous properties by defining 'parent systems' and 'substituent perturbations' [3].

A special case are those molecules which are too unstable to survive storage under normal laboratory conditions. Usually, they are detected, identified and characterized by their molecular state 'fingerprints'. Among the multitude of methods available from the physical armory, photoelectron spectroscopic ionization patterns [4,5] and electron spin resonance signal multiplets [6] may serve as examples for well-developed tools, which are accessible to the preparative chemist and which provide information on either the energy differences between or the structure of individual molecular states [2].

Contributions along these lines and over the years from our group - 65 out of a total of about 250 publications on synthesis and properties of main group element molecules concern organosilicon compounds - are summarized in Figure 1.

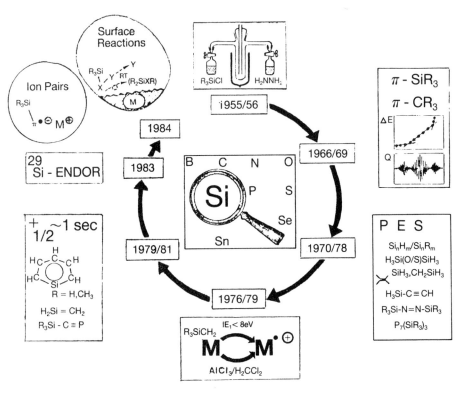

Figure 1 — Contributions to organosilicon chemistry from 1955 to 1984.

Starting out with the gasphase hydrazinolysis of methyl-
chlorosilanes [7] main areas of interest have been:
———— The elucidation of organosilicon substituent effects relative to their carbon analogues by π system perturbation
studies [8].
Synthesis of well over 130 compounds and measurement of
ionization and excitation energies as well as half-wave reduction potentials or radical ion coupling constants hel-
ped to establish especially the powerful electron donor
effect of R_3SiCH_2-groups in radical cations and the acceptor properties of R_3Si-substituents in radical anions
(Figure 2).
———— Photoelectron Spectra and their assignment by quantum chemical calculations [9]:
The ionization patterns i.e. the radical cation state sequences have been rationalized by models based on SiSi/
SiSi bond interactions, lone pair $n_E \rightarrow Si$ delocalization or

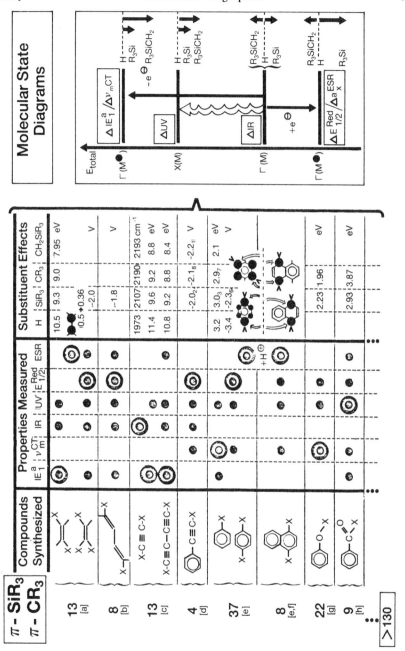

Figure 2 — Substituent effects of R_3SiCH_2- and R_3Si-groups on individual molecular states of π systems ($\odot$: measurements quoted in right-hand column) [8].

H_3Si/π hyperconjugation. The blue bis(trimethylsilyl) azo compound exhibits a low first ionization potential of only 7.1 eV and the $(R_3Si)_3P_7$ cage proves to be the isoelectronic analogue of P_4S_3.

—— Organosilicon Radical Cations [10]:
The selective and oxygen-free one-electron transfer system $AlCl_3/H_2CCl_2$ can oxidize all organosilicon (as well as other main group element) compounds with first vertical (PES) ionization energies below 8 eV. For the numerous radical cations generated spin distributions including [29]Si, configurations and conformations, dynamic processes and reactions have been investigated.

—— Short-Lived Organosilicon Molecules [11]:
A newly designed photoelectron spectrometer with an electron bombardment heated inlet allows to thermolyze compounds up to 1500 K in 2 cm distance from the target chamber. With propene, hexafluoro o-xylene or trimethylchlorosilane as advantageous leaving groups e.g. silatoluene, silabenzene, silaethylene or trimethylsilyl methylidene phosphane could be detected and their ionization patterns assigned by partly rigorous quantum chemical calculations.

Presently, the Frankfurt Group pursues both

—— the optimization of gasphase reactions using PE spectroscopic real-time analysis to generate short-lived molecules [4] and to investigate surface reactions e.g. on silicon, silver or Raney nickel as well as

—— electron transfer studies applying cyclic voltammetry [12] in combination with ESR/ENDOR spectroscopy including the [29]Si ENDOR technique developed in 1983 [13] to investigate radical ions and especially radical ion pairs [14].

Some recent results concerning organosilicon compounds are reported here.

GASPHASE REACTIONS I: SILABENZENE REAFFIRMED

Silabenzene, the 6 π electron perimeter with SiC multiple bonds, has eluded definite prove of its predicted existence until 1979, when some of its PE spectroscopic fingerprints could be detected in a mixture with propene produced by 1000 K-pyrolysis of 5-allyl-5-sila-hexa-1,3-diene [4]:

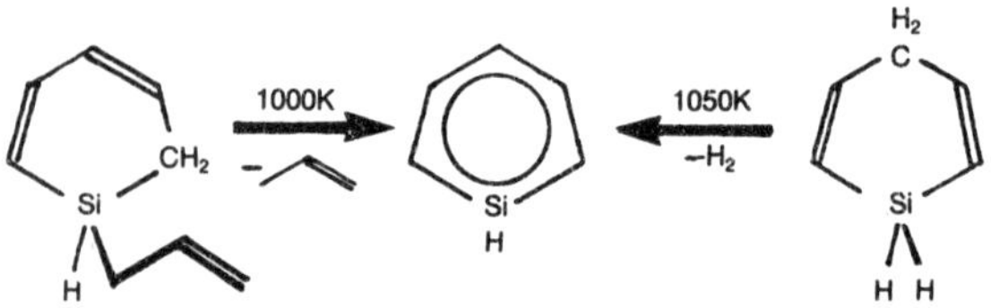

Still lacking, however, have been both a PE spectrum of C_5SiH_6, which unambiguously allows to observe all valence ioni-

zation states within the measurement region and, for its unequivocal assignment, an ab initio SCF calculation employing a basis set of at least double ξ quality. Heating 1-sila-2,5-cyclo-
hexadiene to 1050 K in a PE spectroscopically monitored flow
system [15] , H_2 is expelled as a leaving molecule (1), which
due to its low ionization cross section remains invisible under
the recorded band pattern (Figure 3).

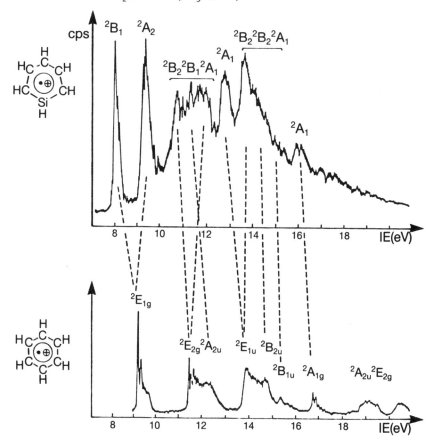

Figure 3 — He(I) PE spectrum of silabenzene at 1050 K and radical cation state comparison
with benzene.

Based both on the PES assignment for benzene - seemingly
settled after much controversy [15] - and on the ab initio SCF
calculation for C_5SiH_6 performed by P. Rosmus - covering its 50
electrons with 76 groups of Gaussian orbitals and yielding a
total energy of - 481.1428 a.u. [15] - some selected details

shall illustrate the 1:1 correspondence between the radical cation state sequences of the two molecules. Following the assigned ionization patterns of Figure 3, two dominant features are recognized: the overall shift to lower ionization energies from C_6H_6 to C_5SiH_6 due to the decreasing effective nuclear charge C > Si and the removal of all degeneracies due to the lowering of the skeletal symmetry from D_{6h} to C_{2v}. Thus the Jahn/Teller broadening of the first PES band of benzene disappears, the π ground state of $C_5SiH_6\cdot^{\oplus}$, $\tilde{X}(^2B_1)$, is lowered by 1.14 eV to 8.11 eV and its second "excited π state, $\tilde{A}(^2A_2)$, moved 1.35 eV apart to 9.46 eV. The third π radical cation state, buried in the ionization hill around 11 eV to 12 eV, presumably [15] has to be located at 11.36 eV. These 3 π-ionization energies of C_5SiH_6 fit perfectly into the heterobenzene π perturbation correlation proposed by E. Heilbronner et al. [16] at the silicon atom ionization potential, $IE_1(Si) = 8.15$ eV:

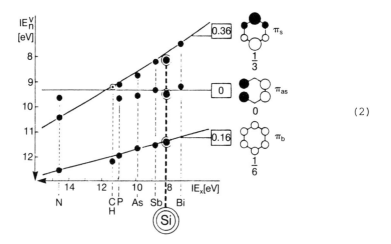

$$(2)$$

The regression lines of diagram (2) exhibit slopes proportional to the squared coefficients $c^2_{j\mu}$ (▭) at the substitution center and have been used already in 1977 by H. Bock to correctly predict the π ionization pattern of C_5SiH_6 [2], providing helpful hints in the search for the unstable molecule. Above all, correlation (2) proves beyond doubt that silabenzene indeed bears the right name.

GASPHASE REACTIONS II: WAYS TO GENERATE DICHLORO SILYLENE ON SILICON SURFACES

The ionization pattern of SiF_2 has been published in 1977 by T.P. Fehlner and D.W. Turner as well as N.P.C. Westwood [17],

who generated the reactive triatomic molecule by either passing
SiF_4 over Si at 1400 K or by heating a mixture of Si and CaF_2 to
1450 K. Also $SiCl_2$ and $SiBr_2$ have been produced in a high-tempe-
rature reactor nozzle system by reaction of Si_2Cl_6 or $SiBr_4$ with
Si at 1470 K and their structure determined by electron diffraction [18]. But, whereas the PE spectra of other group IVb diha-
lides like $SnCl_2$ [19] are reported in the literature, no such
entry is revealed by CAS ON LINE search for $SiCl_2$ [20], a compound of interest as an intermediate in numerous insertion reactions [21].

Applying the available 'state of the art' equipment - a
high-performance PE spectrometer LEYBOLD HERAEUS UPG 200 equip-
ped with an electron bombardment augmented heated inlet system
[22] and connected on line to a VAX 11/750 computer - B. Solouki
from our group succeeded in optimizing by PES real-time analysis
[4] the following reactions of chlorosilane derivatives with
silicon powder dispersed in quartz wool and heated to 1450 K [22]:

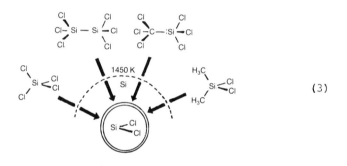

(3)

For example , if trichloromethyl trichlorosilane is passed
over silicon powder, the following changes in the ionization
pattern are observed with increasing temperature (Figure 4):
Above 1000 K the $Cl_3C-SiCl_3$ starts to decompose and at 1250 K
the PE spectrum of pure $SiCl_4$ is recorded. No trace of any
chlorocarbon derivative can be detected; as tentative explanations either formation of $(SiC)_\infty$ within or of a $(CCl_2)_\infty$ coating
outside the oven zone are offered. Except for traces of HCl
especially in the pyrolysis of dimethyldichlorosilane (3), no
side products showed up in the ionization patterns recorded.

All reactions (3) yielded at 1450 K identical PE spectra,
exhibiting each four main bands at 10.3 eV, 11.9 eV to 12.3 eV,
around 13 eV and at 14.0 eV in an approximate intensity ratio
of (0.5):2:2:1. The assignment by Koopmans correlation, $IE_n^V = -\varepsilon_J^{MNDO}$, with MNDO eigenvalues suggests a plausible radical cation state sequence (Figure 4). Accordingly, the dominant con-
tribution in the ground state of $SiCl_2 \cdot^\oplus$ should be a largely
$3s_{Si}$-type silicon lone pair, which also possibly accounts for a
small ionization cross section [23] and thus would explain the

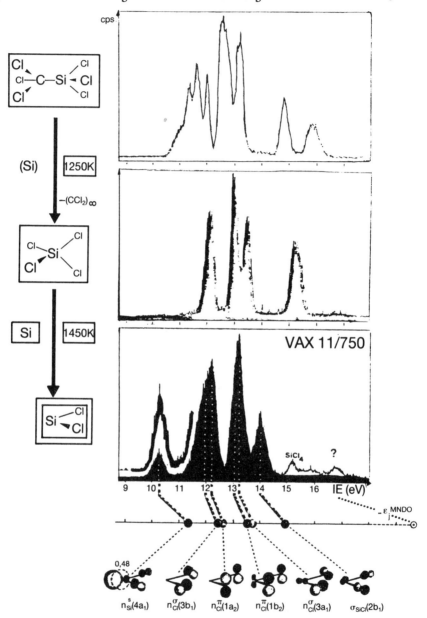

Figure 4 — PE spectroscopically monitored pyrolysis of Cl_3C-$SiCl_3$ on silicon [22] : formation of $SiCl_4$ and of $SiCl_2$ are completed at 1250 K and 1450 K, respectively. The approximate assignment of the $SiCl_2$ ionization pattern is based on a Koopmans correlation with MNDO eigenvalues (see text).

rather low intensity of the PES band at 10.3 eV. The next 4 $M^{\cdot\oplus}$
states are represented by in-plane and out-of-plane chlorine
lone pairs, and the sixth at 14.0 eV by an antibonding combination of σ_{SiCl} bonds. Within the He(I) measurement region only
one more - and again largely $3s_{Si}$ - $M^{\cdot\oplus}$ state is predicted
(ε_7^{MNDO} = 19.7 eV), which could correspond to the 'question-mar-
ked' hump around 18 eV; the other one at 15.2 eV is probably due
to a small $SiCl_4$ contamination (Figure 4). The above assignment
is supported by comparison with the PE spectrum of $SnCl_2$ [19],
which exhibits an analogous 1:2:2:1 band pattern between 10.3 eV
and 12.7 eV with an additional hump at 15.9 eV, assigned to a
(largely $5s_{Sn}$?) state of A_1 symmetry.

The geometry-optimized MNDO calculations yield a structure
exhibiting $d_{SiCl}\sim206$ pm and $\sphericalangle ClSiCl \sim109°$ close to the one determined by gasphase electron diffraction, d_{SiCl} = 208 pm and
$\sphericalangle ClSiCl$ = 103° [18]. The accompanying MNDO charge distribution

(4)

+ 0.95   −0.47

suggests a highly positive Si center, thereby offering an explanation for the insertion activity of $SiCl_2$ into many main group
element bonds [21].

Attempts to generate and to PE spectroscopically identify
other silylenes SiX_2 in gasphase flow systems are under way
[22].

GASPHASE REACTIONS III: ROOM TEMPERATURE NITROGEN SPLIT-OFF FROM ORGANOSILICON COMPOUNDS

In other surface reactions presently studied [24], the solid material placed in the flow system acts as heterogeneous
catalyst. A special activity is exhibited by air-inflammatory
Raney nickel which e.g. already at room temperature dehydroge-
nates isopropanol to acetone [24] or splits off N_2 from diazo
acetic acid methylester to stereoselectively form the 'dimeric'
maleic acid dimethylester [24].

Also organosilicon compounds with N_2-containing groups like
trimethylsilyl diazomethane or the thermally extremely stable
trimethylsilyl azide, on contact with Raney nickel [24] eliminate
N_2 at room temperature (Figure 5). The heterogeneous catalysis
on the vast nickel surface may be rationalized in the following
way:

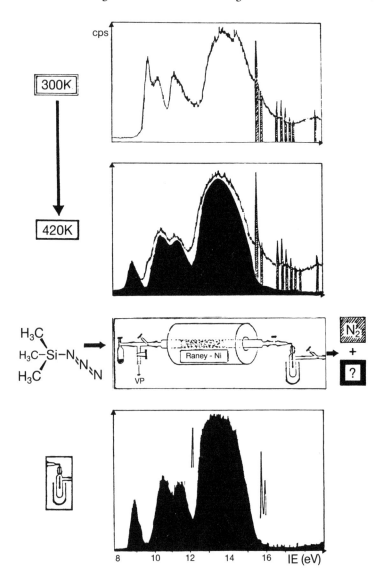

Figure 5 — He(I) PE spectra of $(H_3C)_3SiN_3$ after passing a 5 cm long zone of Raney nickel at room temperature and at 420 K. The trapped reaction mixture on fractionate distillation into the PE spectrometer yields the PE spectrum of a structurally unidentified organosilicon compound presumably containing $N-Si(CH_3)_3$ units (see text).

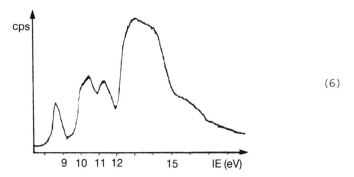

(5)

As concerns the structure of the main decomposition product, which also has been trapped at 200 K from the flow system and reidentified by its ionization pattern on fractionate evaporation into the PE spectrometer (Figure 5), PE spectroscopy itself is not of much help: according to a useful rule of thumb [5], within the He(I) measurement region, all n_p-type element and 1_s-type hydrogen electrons have to be counted together and subdivided by 2 applying the Pauli principle. Thus for e.g. bis-(trimethylsilyl)amine, $((H_3C)_3Si)_2NH$, 19 ionization bands are expected, of whom most overlap in the 12 eV - 16 eV region, and only 3 are separated in front of it exhibiting maxima at 8.7 eV, 10.3 eV and 11.3 eV [25]:

(6)

Comparison of the ionization patterns of both the R_3SiN_3 main decomposition product formed on Raney nickel (Figure 5) and $(R_3Si)_2NH$ (6) exhibits some similarity and leads to the assumption that the yet structurally unidentified product should contain R_3SiN subunits. Attempts to optimize the N_2 elimination from R_3SiN_3 on Raney nickel in a preparative scale and to gain further information on the compound formed by MS and NMR analysis, are presently pursued. Also included in the investigation is phenyl triazido silane, $H_5C_6-Si(N_3)_3$, which on heating in a gasphase flow system at 10^{-3} mbar pressure exhibits an unexpectedly high thermal stability: N_2 split-off can be observed only at temperatures above 950 K [26], and - by comparison with the PE spectrum of H_5C_6-NC as well as geometry-optimized MNDO calculation - evidence for the formation of H_5C_6-NSi, a compound with a $Si≡N$ triple bond, is obtained.

ELECTRON TRANSFER I: ^{29}Si-ENDOR

Even high-resolution ESR spectra of organosilicon radical cations
[6] hardly can cope with the sometimes enormous number of signals and their intensity differences especially in fluctuating
species. Thus for 1,4,5,8-tetrakis(trimethylsilyl)-$\Delta^{4a(8a)}$-oc-
talin radical cation (8) due to 4 sets of each 4,4,4 and 36
equivalent ^{1}H nuclei and 1 ^{29}Si nucleus (I_{Si} = 1/2, natural
abundance 4.70 %) a total of

$$\text{ESR:} \quad \prod_{N=1}^{N} (2n_N I_N + 1) = 5\times5\times5\times37 + 2(5\times5\times5\times37) = 13875\,! \tag{7}$$

signals are predicted, and for the center line of its ESR spec-
rum (Figure 6) alone 925 signals, of which about 50 become dis-
tinguishable at 180 K [27]:

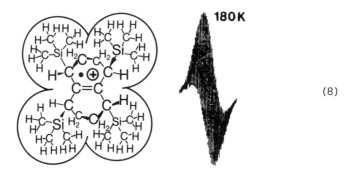

(8)

Application of the 'Electron Nuclear DOuble Resonance'-techni-
que [28] reduces the ESR complexity to just

$$\text{ENDOR:} \quad 2N \quad = \quad 2\times5 \quad = 10 \tag{9}$$

lines i.e. 1 pair of lines for each set of equivalent nuclei N,
which according to the ENDOR resonance condition

$$^{\nu}\text{ENDOR} = \left| \nu_N \pm a_N/2 \right| \tag{10}$$

are centered either around the nuclear frequency ν_N with distance a_N, or - for $a_N/2 > \nu_N$ - around $a_N/2$ with distance $2\nu_N$.
   In order to take advantage of the tremendous gain in information provided by ENDOR [28] - an NMR experiment in the ESR
mode and of special importance for the investigation of complex
systems e.g. in biochemistry [28] - also for organosilicon radicals like (8), B. Hierholzer from our group in a joint pro-
ject with H. Kurreck and M. Lubitz attempted to measure ^{29}Si
ENDOR signals for the first time [28].
   In a preceding feasibility study for the tetrasilyl octaline
radical cation (8) as generated by the selective oxygen-free

oxidation system AlCl$_3$ in H$_2$CCl$_2$ [6], the necessary measurement
conditions have been approximated [29]. The essential parameter
to be optimized is the rotation correlation time τ_{Rot}, the square of which is proportional to the ratio between nuclear (W_I)
and spin/lattice (W_e) relaxation rates. An estimate can be a-
chieved based either on the Debye/Einstein equation or on the
Freed relaxation theory [29]: Both, from the effective volume
V_{eff} and the temperature dependance of the H$_2$CCl$_2$ viscosity, as
well as from empirical (B) and calculated (TrA_0^2) anisotropy factors together with the Si spin population ρ_{Si}, the same range
for τ_{Rot} is predicted [29]:

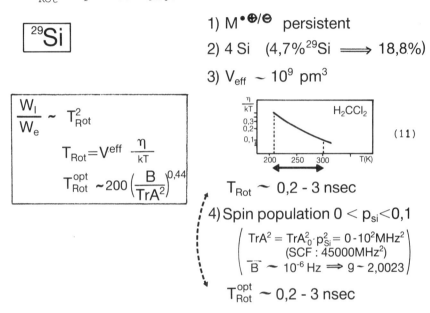

1) $M^{\bullet \oplus / \ominus}$ persistent

2) 4 Si $\quad (4,7\%\,^{29}Si \Longrightarrow 18,8\%)$

3) $V_{eff} \sim 10^9$ pm^3

$$\frac{W_I}{W_e} \sim T_{Rot}^2$$

$$T_{Rot} = V^{eff}\,\frac{\eta}{kT}$$

$$T_{Rot}^{opt} \sim 200\left(\frac{B}{TrA^2}\right)^{0,44}$$

$$T_{Rot} \sim 0,2 - 3 \text{ nsec}$$

4) Spin population $0 < p_{si} < 0,1$

$$\left(\begin{array}{c} TrA^2 = TrA_0^2 \cdot p_{Si}^2 = 0 - 10^2 MHz^2 \\ (SCF : 45000MHz^2) \\ \bar{B} \sim 10^{-6} Hz \Longrightarrow g \sim 2,0023 \end{array}\right)$$

$$T_{Rot}^{opt} \sim 0,2 - 3 \text{ nsec}$$

(11)

Therefore, if the radical ion is persistent and contains
several equivalent Si centers (11), there should be a 'measurement window' in the multiparameter space of radical concentration, temperature, solvent viscosity, spin density etc. to detect ^{29}Si signals in natural abundance [29].

In the tetrasilyl octaline radical cation example chosen
here [29], NMR saturation in both the ^{1}H center and the ^{29}Si
satellite lines (Figure 6) yielded ^{1}H and ^{1}H + ^{29}Si ENDOR spectra (Figure 6), which allow to read off directly the correspond-
ing ^{1}H and ^{29}Si coupling constants a_{1_H} and $a_{29_{Si}}$, respectively.

(12)

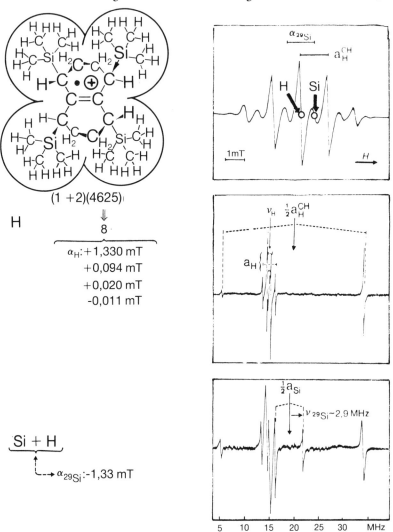

Figure 6 — ESR spectrum of 1,4,5,8-tetrakis(trimethylsilyl)-$\Delta^{4a(8a)}$-octalin radical cation indicating the points of ESR saturation (→0) for the additional NMR sweep to record the ^{1}H + ^{29}Si ENDOR spectra (cf. text).

Performing in addition GENERAL TRIPLE [28] experiments [29], also the signs of the coupling constants could be determined, providing complete information on the ESR multiplet signal pattern of $((H_3C)_3Si)_4C_{10}H_{12}\cdot^{\oplus}$ (12), from which other $M\cdot^{\oplus}$ properties

like the spin population at the individual centers or the conformation in solution (12) can be deduced [29].

The ^{29}Si ENDOR results, completed by temperature-dependent
measurements [29], not only substantiate the estimates of the
preceding feasibility study (11) but also demonstrate as another
advantage of the successful $AlCl_3/H_2CCl_2$ oxidation system its 'EN-
DOR capability'. In addition, organosilicon radical anions generated by alkali metal reduction in ether solvents like

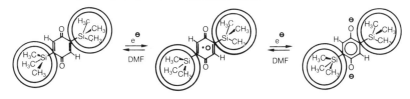

(13)

can be subjected to ENDOR studies and in most cases ^{29}Si line
pairs observed [29,30]. Furthermore, ENDOR spectroscopy proves
to be a most valuable tool to study ion pair interactions (cf.
chapter 'Electron Transfer III').

ELECTRON TRANSFER II: THE REDUCTION OF 2,5-BIS(TRIMETHYL-SILYL)-p-BENZOQUINONE

The two-step electron transfer to the orange-coloured 2,5-
bis(trimethylsilyl)-p-benzoquinone [31]

via its ESR spectroscopically characterized [32] semiquinone radical anion (cf. Figures 8 and 9) to the corresponding hydroquinone dianion is reinvestigated in DMF [33] and yields the
following results:
—— The cyclovoltammogram (Figure 7); counter cation ($H_9C_4)_4N^{\oplus}$)
   shows both consecutive reductions at -0.46 V and -1.37 V
   vs. SCE to be completely reversible.
—— Comparison with the reversible half-wave reduction potentials of the parent p-quinone and of its 2,5-di(tert. butyl) derivative, determined under the same aprotic conditions [33,34]:

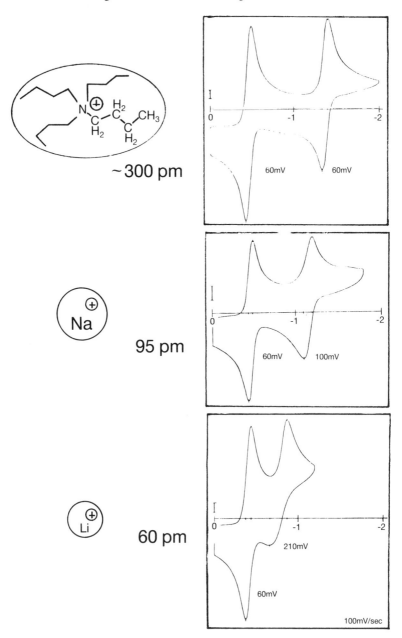

Figure 7 — Cyclovoltammograms for the two-step reduction of 2,5-bis(trimethylsilyl)-p-benzoquinone in DMF at 300 K at a GCE vs. SCE: the effect of counter cations $(H_9C_4)_4N^\oplus$, $Na^\oplus$ and $Li^\oplus$ added as perchlorates (see text).

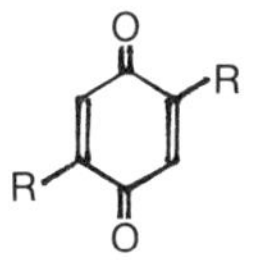

	R		$(H_3C)_3C$	$(H_3C)_3Si$	H	
$E_{1/2}^{Red\,I}$	(V)		-0.59	-0.46	-0.40	(15)
$E_{1/2}^{Red\,II}$	(V)		-1.53	-1.37	-1.30	

demonstrates that the R_3Si substituents - in spite of the lower effective nuclear charge of $Si < C$ - act as weaker do-nors than the R_3C group towards the electron-acceptor hete-ro-π-system containing 2 oxygen centers (cf. Figure 2). Insertion of the first half-wave reduction potential of 2,5-bis(trimethylsilyl)-p-benzoquinone into a electrochemical series established under the same aprotic measurement con-ditions [34]:

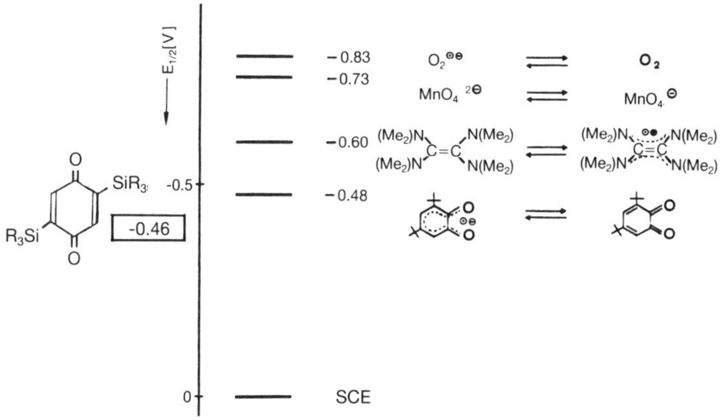

(16)

suggests as 'off the beaten track' methods to selectively generate its semiquinone radical anion e.g. the electron transfer from hyperoxide ion $O_2{}^{\cdot\ominus}$ or from tetrakis(dime-thylamino)ethylene. Both worked: O_2 is irreversibly evol-ved from the solution with KO_2 in THF accomplished by [2.2.2] cryptand addition, and the almost colourless so-lution with $(R_2N)_2C=C(NR_2)$ turned intensively yellow due to formation of the radical cation/radical anion pair:

(17)

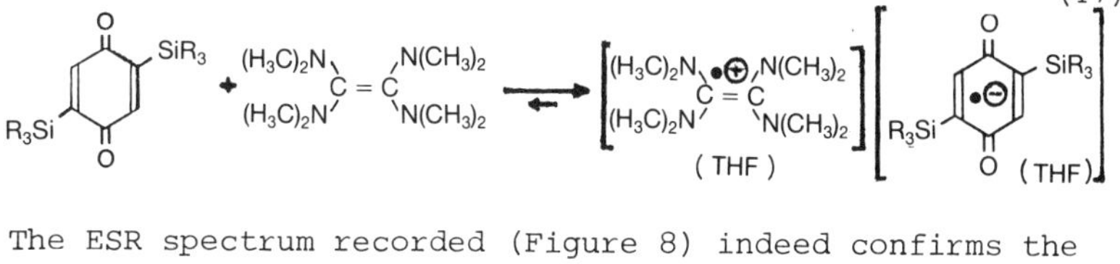

The ESR spectrum recorded (Figure 8) indeed confirms the thermodynamical prediction (16): It shows the overlapping signal patterns both of tetrakis (dimethylamino)ethylene radical cation [35], in which well over 300 lines can be distinguished, and - with higher intensity - the characte-ristic ^{1}H triplet of 2,5-bis(trimethylsilyl)-p-semiquinone radical anion [33].

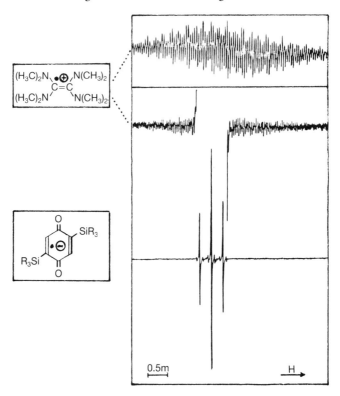

Figure 8 — ESR spectrum of the tetrakis(dimethylamino)ethylene radical cation + 2,5-bis(trimethylsilyl)-p-benzosemiquinone radical anion pair in THF. For comparison, the ESR signal pattern of $(R_2N)_2C=C(NR_2)^{.\oplus}$ is shown above (.....); for that of $(R_3Si)_2C_6H_2O_2^{.\ominus}$ cf. Figure 9 (see text).

Another surprising result is obtained from cyclovoltammograms recorded in DMF solutions containing either $Na^{\oplus}ClO_4^{\ominus}$ or $Li^{\oplus}ClO_4^{\ominus}$ as conductivity salt (Figure 7). Relative to $R_4N^{\oplus}$ as counter cation, especially the second potential is lowered by 0.23 V to -1.14 V for $Na^{\oplus}$ and by 0.60 V to -0.77 V for $Li^{\oplus}$. Simultaneously, reversibility decreases from the 'ideal' peak-to-peak distance $\Delta E = 60$ mV for $R_4N^{\oplus}$ to 100 mV for $Na^{\oplus}$ and 210 mV for $Li^{\oplus}$. Both changes indicate, that ion pair interactions take place: Whereas the bulky tetrabutylammonium cation - with its electrochemically effective radius of about 300 pm [34] and with its largely $N^{\oplus}$ localized charge packed in 'hydrocarbon wad' - may be considered an independent and inert moiety, the formation of radical contact ion pairs:

$$(R \cdot \ominus)_{DMF} + (M^{\oplus})_{DMF} \rightleftharpoons [R \cdot {}^{\ominus}M^{\oplus}]_{DMF} \qquad (18)$$

in general should be favored by decreasing counter cation radius. Assuming either $p = 1$ or 2 cations becoming more closely attached to the semiquinone radical anion in the first reduction step, by inserting the observed potential difference $\Delta E_{1/2}$ and the cation concentration $c_{M\oplus}$ into the modified Nernst equation

$$\Delta E_{1/2} = \frac{RT}{F} \ln (K_{ass} - p \cdot c_{M\oplus}) \qquad (19)$$

association constants K_{ass} in the range of 35 to 70 $[1/mole]^n$ for $Li^{\oplus}$ are estimated [33].
The cyclovoltammetric data for the two-step electron transfer to 2,5-bis(trimethylsilyl)-p-benzoquinone in the presence of $Li^{\oplus}$ as well as coinciding studies on ion pair formation of other (silicon-free) main group element compounds [33,34,36] stimulated the following ESR/ENDOR investigation.

ELECTRON TRANSFER III: ION PAIR FORMATION OF TRIMETHYLSILYL-SUBSTITUTED QUINONES

The cyclovoltammetrically detected large shifts especially of the second reduction potential of 2,5-bis(trimethylsilyl)-p-quinone (Figure 7) suggested contact ion pair formation $[M \cdot {}^{\ominus}Li_n^{\oplus}]$, which for semiquinone radical anion and its alkyl derivatives has been first observed due to varying ESR line shapes by E.A.C. Lucken back in 1964 [37]. In the mean-time, numerous ESR [38] and ENDOR [28,39] investigations applying this 'molecular finger-print' signal technique have shed more light on the complex attraction phenomena between ions in organic solvents mostly under aprotic conditions.
    In spite of an intense literature search, no report on ion pair formation involving an organosilicon radical anion could be found [33]. The following ESR/ENDOR results are obtained for mono- and 2,5-bis(trimethylsilyl)-substituted semiquinone radical anions in the presence of the counter-cations $Li^{\oplus}$, $K^{\oplus}$ as well as $Tl(C_6H_5)_2^{\oplus}$:
—— The changing ESR multiplet signal pattern of 2,5-bis(trimethylsilyl)-p-benzoquinone radical anion in THF solution containing varying concentrations of $Li^{\oplus}$ counter cations (Figure 9) clearly demonstrate the formation of double and triple ion radical species. The 'naked' 2,5-bis(trimethylsilyl)-p-benzoquinone radical anion, generated by reduction with lithium metal in the presence of an excess of [2.2.1] cryptand exhibits - besides ^{13}C and ^{29}Si satellites of low intensity a ^1H triplet for the 2 equivalent ring protons. On reduction with lithium metal without addition of a complexing

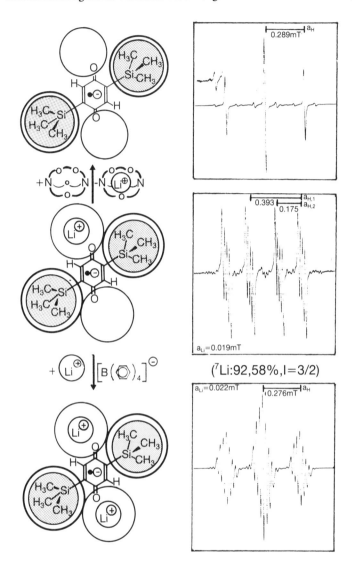

Figure 9 — ESR spectra of 'naked' 2,5-bis(trimethylsilyl)-p-benzoquinone radical anion, generated by lithium metal reduction in the presence of [2.1.2] cryptand, the radical $[M \cdot {}^\ominus Li^\oplus] \cdot$ formed directly on reduction with lithium metal and the triple radical cation $[M \cdot {}^\ominus (Li^\oplus)_2] \cdot {}^\oplus$ prepared by adding $Li^\oplus [B(C_6H_5)_4]^\ominus$ i.e. by increasing the $Li^\oplus$ concentration in the THF solution.

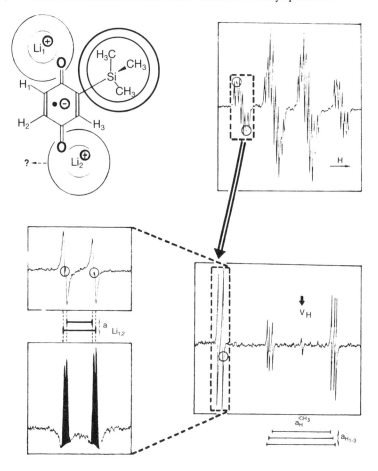

Figure 10 — ESR and ENDOR spectra of the contact radical cation formed between trimethyl-silyl-p-benzoquinone radical anion and 2 Li$^\oplus$ cations in THF.

agent, the radical ion pair [M·$^\ominus$Li$^\oplus$]· is formed, showing a ^{1}H doublet of doublets for the no longer equivalent quinone hydrogens with each line splitted into a quartet due to the additional coupling by one ^{7}Li$^\oplus$ (I = 3/2, natural abundance 92.58 %) fixed into one of the two equivalent positions near the carbonyl subunits. Increasing the Li$^\oplus$ concentration in the THF solution by addition of lithium tetraphenyl borate leads to the radical triple cation [M·$^\ominus$(Li$^\oplus$)$_2$]·$^\oplus$ as deduced from the triplet for the again equivalent hydrogens splitted each into a septet due to the coupling of two ^{7}Li$^\oplus$ cations now covering both available and equivalent positions near the carbonyl subunits of the quinone (Figure 9).

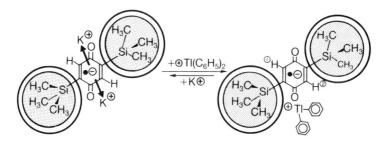

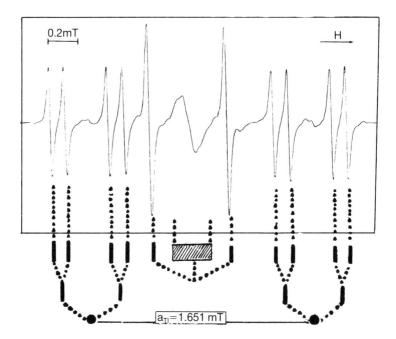

Figure 11 — The temperature-dependent equilibrium between $K^{\oplus}$ and $Tl(C_6H_5)_2^{\oplus}$ cation attachment to the radical anion of 2,5-bis(trimethylsilyl)-p-benzoquinone, recorded at 295 K and exhibiting a fluctuating $K^{\oplus}$ and a fixed $Tl(C_6H_5)_2^{\oplus}$ cation, respectively (see text).

The fixation of the $Li^{\ominus}$ counterions into positions near the carbonyl subunits of the semiquinone radical anion can be further substantiated by the ESR and ENDOR spectra (Figure 10) of trimethylsilyl-p-benzoquinone radical anion in THF solution containing an excess of $Li^{\ominus}$ cations as achieved by addition of $Li^{\ominus}[B(C_6H_5)_4]^{\ominus}$: the 3 doublet ESR signal pattern for the nonequivalent ring-hydrogens is reduced by ENDOR (9) to 3 line pairs; due to the increased resolution also a sig-

nal for the 9 equivalent $(H_3C)_3Si$ hydrogens surfaces. The
additional ^7Li splitting of the ESR ^1H multiplet can either
result from 1 septet or - as indicated by little humps (Figure 10:0) - 2 quartets if the 2 $Li^\oplus$ are fixed into nonequivalent positions. That this is indeed the case can be proven
beyond doubt by the high resolution $^7Li^\oplus$ ENDOR signals which-
especially in the 2nd derivative of the spectrum - clearly
show 2 different ^7Li line pairs (Figure 10).

———— Among the many cases investigated [33], also those with fluctuating cations are observed. An especially convincing example is the $K^\oplus$ vs. $Tl(C_6H_5)_2^\oplus$ equilibrium (Figure 11): At
the temperature of 295 K chosen, the $Tl(C_6H_5)_2^\oplus$ cation is
fixed into a certain position within the vicinity of 2,5-bis
(trimethylsilyl-p-benzoquinone radical anion as detectable
by the ^1H doublets of doublets separated by a rather large
$^{203,205}Tl$ (I = 1/2) splitting. On the other hand, at the
very same temperature the $K^\oplus$ starts moving over the semiquinone surface within the ESR time-scale - as clearly demonstrated by the 'snap-shot' of the changing line pattern (Figure 11: hatched area) with a ^1H doublet of doublets for 2
non-equivalent ring hydrogens beginning to be transformed
into a triplet for 2 equivalent ones.

———— The electron transfer on to quinones, the ion pair formation
and the ion movement over the π plane surface are followed
by extensive MNDO hypersurface calculations. Comparison of
the results for 2,5-dimethyl and 2,5-bis(trimethylsilyl)-p-
benzoquinones shows:

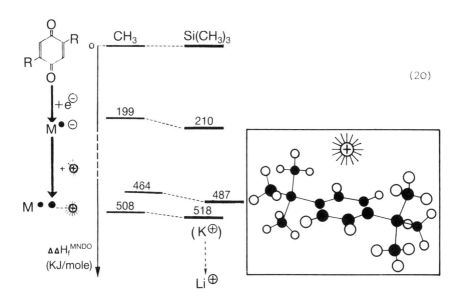

(20)

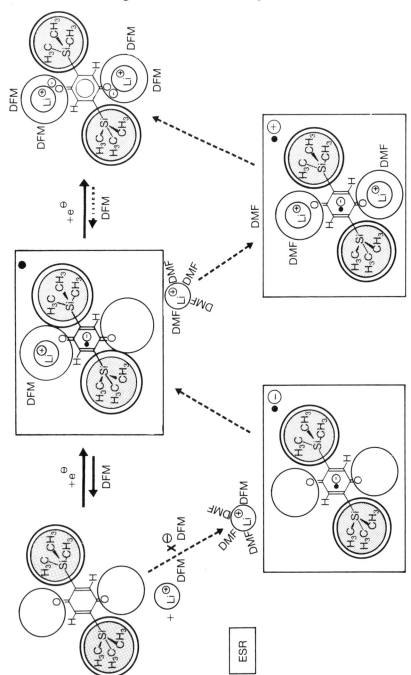

Figure 12 – The reduction of 2,5-bis(trimethylsilyl)-p-benzoquinone simplified (see text).

The largest difference in MNDO heats of formation, $\Delta\Delta$ H$_f^{MNDO}$, is due to the addition of the counter cation - simulated by a 'sparkle' (2O: ⊙) representing approximately K$^\oplus$ in size - to the radical anion formed by electron transfer. Allowing the (K$^\oplus$) 'sparkle' to move, a gradient search program following its pathway of lowest energy - cf. the computerplot of the sparkle ⊕ at the saddle point over 2,5-bis(trimethylsilyl)-p-benzoquinone radical anion (2O) - suggests only small activation barriers of 44 kJ/mole and 31 kJ/mole respectively, with the one for the organosilicon species being smaller. This is in full agreement with the fluctuation of K$^\oplus$ observed ESR spectroscopically at about room temperature (Figure 11). For Li$^\oplus$, exercising a considerably higher effective ionic charge, a stronger ion pair interaction is expected causing its fixation near the carbonyl subunits in the semiquinone molecular plane as convincingly demonstrated by the ESR and ENDOR spectra recorded for its radical ion complexes formed (Figures 9 and 10).

To cut a rather complex story short, both the cyclovoltammetric and the ESR/ENDOR results are summarized in a simplified, self-explicable way (Figure 12) with the 3 radical species detected ESR/ENDOR spectroscopically - the 'naked' radical anion of 2,5-bis(trimethylsilyl)-p-benzoquinone M$^{\bullet\ominus}$, its mono Li$^\oplus$ adduct radical [M$^{\bullet\ominus}$Li$^\oplus$]$^\bullet$ and its 'triple ion complex'-radical cation [M$^{\bullet\ominus}$(Li$^\oplus$)$_2$]$^{\bullet\oplus}$ - in the center, the multifaceted network of equilibria is indicated through which the seemingly straightforward two-step reduction of a quinone via its semiquinone to the hydroquinone dianion may have to pass.

The example for ion pair formation by organosilicon compounds presented here will, hopefully, stimulate other investigations and also might have technical implications.

Thanks are expressed expecially to the coworkers mentioned, who - presently working in other areas of main group chemistry - on short notice switched to organosilicon chemistry and produced in less than half a year the results reported here. Our research has been and is generously funded by Land Hessen, Deutsche Forschungsgemeinschaft and Fonds der Chemischen Industrie as well as by the A. Messer, M. Buchner and H. Schlosser Foundations.

REFERENCES

1.   11th Essay on Molecular Properties and Models. For the
      preceding 10th on 'Molecular State Fingerprints and Semi-
      empirical Hypersurface Calculations - Useful Correlations
      to Track Short-Lived Molecules' cf. H. Bock, R. Dammel
      and B. Roth, in 'Rings, Clusters and Polymers of the Main
      Group Elements' (Edit. A. H. Cowley), ACS Symposium Se-
      ries, Vol. 232, American Chemical Society, Washington DC.,
      1983, p. 139-165.

2.   H. Bock, Angew. Chem. 1977, 89, 613-637, Angew. Chem. Int.
      Ed. Engl. 1977, 16, 613-637, and literature cited.

3.   Cf. e.g. E. Heilbronner and H. Bock 'The HMO Model' Vol. 1,
      Verlag Chemie Weinheim 1969, Hirokawa Tokyo 1973, Wiley &
      Sons London 1976 and Kirin Univ. Press (China) 1982, Chap-
      ters 6 and 7.

4.   Cf. e.g. H. Bock and B. Solouki, Angew. Chem. 1981, 93,
      425-442; Angew. Chem. Int. Ed. Engl. 1981, 20, 427-444
      with 142 literature quotations.

5.   See also H. Bock and B.G. Ramsey, Angew. Chem. 1973, 85,
      773-792; Angew. Chem. Int. Ed. Engl. 1973, 12, 734-752 and
      literature cited.

6.   Cf. e. g. H. Bock and W. Kaim, Acc. Chem. Res. 1982, 15,
      9-17 and literature cited.

7.   H. Bock, Z. Naturforsch. 1962, 17b, 423; cf. Ph. D. Thesis
      H. Bock, University of Munich 1958.

8.   In cooperation with H. Alt, H. Seidl and U. Krynitz as well
      as F. Gerson and E. Heilbronner. For a summary cf. H. Bock,
      Jahrbuch Akademie der Wissenschaften Göttingen 1969, 13-25.
      For details in Figure 2 cf. [a] H. Bock and H. Seidl, J.
      organomet. Chem. 1968, 13, 807. [b] H. Bock and H. Seidl,
      J. Amer. Chem. Soc. 1968, 90, 5694, as well as J. Kroner
      and H. Bock, Theor. Chim. Acta 1968, 12, 214. [c] H. Bock
      and H. Seidl, J. Chem. Soc. B 1968, 1158. [d] H. Bock and
      H. Alt, Chem. Ber. 1970, 103, 1784. [e] H. Bock, H. Seidl
      and M. Fochler, Chem. Ber. 1968, 101, 2815; H. Bock and
      H. Alt, J. Amer. Chem. Soc. 1970, 92, 1569; F. Gerson,
      J. Heinzer, H. Bock and H. Seidl, Helv. Chim. Acta 1968,
      51, 707 as well as H. Alt, E.R. Francke and H. Bock, An-
      gew. Chem. 1969, 14, 538; Angew. Chem. Int. Ed. Engl. 1969,
      7, 525. [f] Cf. [e] as well as H. Bock and H. Alt, Chem.
      Ber. 1969, 102, 1534. [g] H. Bock and H. Alt, J. organomet.
      Chem. 1968, 13, 103 as well as H. Alt and H. Bock, Tetrah.
      1971, 27, 4695. [h] H. Bock, H. Alt and H. Seidl, J. Amer.
      Chem. Soc. 1969, 91, 355 as well as F. Gerson. U. Krynitz
      and H. Bock, Helv. Chim. Acta 1969, 2, 2512.

9.   In cooperation with W. Enßlin, P. Mollère, B. Solouki and
      K. Wittel as well as G. Becker, F. Fehér, G. Fritz and N.
      Wiberg. For a summary cf. [5].

10. In cooperation with W. Kaim, M. Kira, and G. Brähler as well as G. Fritz with whom the first organosilicon radical cation has been generated) H. Sakurai, R. West and N. Wiberg. For a summary cf. [6].

11. In cooperation with B. Solouki and P. Rosmus as well as R. Appel, T. Barton and G. Maier. For a summary cf. [4]. For $H_2Si=CH_2$ cf. P. Rosmus, H. Bock, B. Solouki, G. Maier and G. Mihm, Angew. Chem. 1981, 93, 616; Angew. Chem. Int. Ed. Engl. 1981, 20, 598, and for $(H_3C)_3Si-C\equiv P$ cf. B. Solouki, H. Bock, R. Appel, A. Westerhaus, G. Becker and G. Uhl, Chem. Ber. 1982, 115, 3747.

12. Cf. e.g. H. Bock and D. Jaculi, Angew. Chem. 96, 74 (1984); Angew. Chem. Int. Ed. Engl. 23, 57 (1984).

13. H. Bock, B. Hierholzer, H. Kurreck and W. Lubitz, Angew. Chem. Int. Ed. Engl. 22, 787 (1983).

14. Cf. e.g. H. Bock, B. Hierholzer, F. Vögtle and G. Hollmann, Angew. Chem. 96, 74 (1984); Angew. Chem. Int. Ed. Engl. 23, 57, (1984).

15. The publication by H. Bock, P. Rosmus, B. Solouki and G. Maier, J. Organomet. Chem. 1984, in print, is dedicated to Professor Kumada on occasion of his 70th birthday, and quotes 32 literature references.

16. C. Batich, E. Heilbronner, V. Hornung, A.J. Ashe III, D.T. Clark, U.T. Cobley, D. Kilcast and I Scanlan, J. Amer. Chem. Soc. 1973, 95, 928. For C_5BiH_6 cf. J. Bastide, E. Heilbronner, J.P. Maier and A.J. Ashe III, Tetrah. Letters 1976, 411.

17. T.P. Fehlner and D.W. Turner, Inorg. Chem. 1974, 13, 754 as well as N.P.C. Westwood, Chem. Phys. Letters 1974, 25, 558.

18. I. Hargittai, G. Schultz, J. Tremmel, N.D. Kagramanov, A.K. Maltsev and O.M. Nefedov, J. Amer. Chem. Soc. 1983, 105, 2895 and literature cited.

19. Cf. e.g. S. Evans and A.F. Orchard, J. Electron Spectry. Rel. Phen. 1975, 6, 207 or D.H. Harris, M.F. Lappert, J.B. Pedley and G.J. Sharp, JCS Dalton Trans. 1976, 945.

20. The CAS ON LINE literature search is dated July 23rd, 1984.

21. For an early review on generation and reactivity of silylenes see W.H. Atwell and D.R. Weyenberg, Angew. Chem. 1969, 81, 485 - 493, Angew. Chem. Int. Ed. Engl. 1969, 8 469 - 477. Cf. also the abstracts of the VII. International Symposium on Organosilicon Chemistry (Ed. H. Sakurai), Ellis Horwood Lim., Chichester, to be published, and the more recent literature cited there.

22. H. Bock and B. Solouki, unpublished results. For the instrumentation used cf. [4] and Chem. Ber. 1982, 115, 3748.

23. Cf. e.g. J.W. Rabalais 'Principles of Ultraviolet Photoelectron Spectroscopy', Wiley & Sons, New York, p. 139 - 178.

24. H. Bock and H.P. Wolf, unpublished results. Cf. Master Thesis H.P. Wolf, University of Frankfurt 1984.

25. K. Ostoja Starzewski, H. tom Dieck and H. Bock, J. organo-
     met. Chem. 1974, 65(1974).
26. H. Bock and R. Dammel, unpublished results. For PES-moni-
     tored pyrolysis of azides cf. e.g. H. Bock, R. Dammel and
     S. Aygen, J. Amer. Chem. Soc. 1983, 105, 7681 and litera-
     ture cited.
27. H. Bock and W. Kaim, J. Amer. Chem. Soc. 1980, 102, 4437.
28. For a recent summary on ENDOR spectroscopy cf. H. Kurreck,
     B. Kirste and W. Lubitz, Angew. Chem. 1984, 96, 171-193;
     Angew. Chem. Int. Ed. Engl. 1984, 23, 173 - 195 and lite-
     rature cited.
29. H. Bock, B. Hierholzer, H. Kurreck and W. Lubitz, Angew.
     Chem. 1983, 95, 817. Angew. Chem. Int. Ed. Engl. 1983, 22,
     787. For details cf. Angew. Chem. Suppl 1983, 1088 or
     Master Thesis B. Hierholzer, University of Frankfurt 1982
     and literature cited.
30. H. Bock and B. Hierholzer, unpublished results from joint
     projects with H. Balli (University of Fribourg, Switzer-
     land) together with S. Masamune (MIT Boston), E. Hengge
     (University Graz), M. Weidenbruch (University Oldenburg)
     and R.W. West (Madison/Wisconsin).
31. H. Bock and H. Alt, Angew. Chem. 1967, 79, 932; Angew.
     Chem. Int. Ed. Engl. 1967, 6, 941. The half-wave reduction
     potentials have been determined polarographically vs. the
     Hg anode in H3CCN/tetraethyl ammonium iodide; cf. Thesis
     H. Alt, University of Munich 1969.
32. F. Gerson, U. Krynitz and H. Bock, Helv. Chim. Acta 1969,
     52, 2513. The 2,5-bis(trimethylsilyl)-p-semiquinone radi-
     cal anion has been generated electrolytically in DMF/TEAP
     at 300 K; cf. Thesis U. Krynitz, University of Munich
     1969.
33. H. Bock, P. Hänel, H. Herrmann and U. Lechner-Knoblauch,
     unpublished results. The CV measurements are performed in
     DMF/tetrabutylammonium perchlorate at 300 K using a glassy
     carbon electrode vs. SCE. In order to achieve ppm aprotic
     conditions [34], e.g. all operations are carried out un-
     der Ar; cf. Thesis U. Lechner-Knoblauch, University of
     Frankfurt 1984, and literature cited.
34. H. Bock and D. Jaculi, Angew. Chem. 1984, 96, 298; Angew.
     Chem. Int. Ed. Engl. 1984, 23, 305. Cf. Master Thesis D.
     Jaculi, University of Frankfurt 1983, and literature ci-
     ted.
35. K. Kuwata and D.H. Geske, J. Amer. Chem. Soc. 1964, 86,
     2101. Cf. also the summary by N. Wiberg, Angew. Chem. 1968,
     80, 809-822; Angew. Chem. Int. Ed. Engl. 1968, 7, 766-779.
36. Cf. e.g. H. Bock, B. Hierholzer, F. Vögtle and G. Holl-
     mann, Angew. Chem. 1984, 96, 74; Angew. Chem. Int. Ed.
     Engl. 1984, 23, 57.
37. E.A.C. Lucken, J. Chem. Soc. 1964, 4234.

38.   Cf. e.g. N. Hirota, J. Amer. Chem. Soc. 1968, $\underline{90}$ (1968),
       or T.E. Gough and P.R. Hindle, Canad. J. Chem. 1969, $\underline{47}$,
       1698 as well as 3393, and literature cited.

39.   Cf. e.g. W. Lubitz, M. Plato, K. Möbius and R. Biehl, J.
       Phys. Chem. 1979, $\underline{83}$, 3402 and literature cited.

CHAPTER 6

SOME INSIGHTS INTO THE CHEMISTRY OF ORGANOSILICON INTERMEDIATES FROM GAS KINETIC STUDIES

Iain M. T. Davidson Department of Chemistry, The University, Leicester, LE1 7RH, U.K.

We have found it fascinating to pursue gas kinetic studies in organosilicon chemistry because it seems to us that the interplay between different types of intermediate is particularly subtle, leading to intriguing and challenging mechanistic problems. We hope to show in this contribution that kinetic experiments and estimates have now progressed sufficiently to be of some use in resolving these problems.

An apposite example illustrating both the similarities and differences between carbon and silicon chemistry, and the rôle of gas kinetics in rationalising apparent mechanistic dichotomies, is the pyrolysis of hexamethyldisilane. As might be expected by analogy with hydrocarbons, $Me_3SiSiMe_3$ dissociates thermally to produce $Me_3Si\cdot$ radicals. Kumada found that pyrolysis of $Me_3SiSiMe_3$ at high pressure gave a high yield of the isomer, $Me_3SiCH_2Si(Me_2)H$ (I), a result which was very reasonably interpreted [1] in terms of a radical chain reaction involving a unimolecular radical rearrangement. We obtained kinetic data for this isomerization [2], but on the other hand, we found that pyrolysis at low pressure gave 1,1,3,3-tetramethyl-1,3-disiletane (II) and trimethylsilane as the main products [3]. By reinforcing our pyrolysis kinetics [3] with experiments in which radicals were generated photochemically at lower temperatures [4], we were able not only to confirm the mechanism shown in Scheme 1 below, which accounts for the apparently disparate results observed under different conditions, but to measure Arrhenius parameters for the individual elementary reactions (1), (3), (4) and (5). Reactions (1) to (4) are those suggested by Kumada. There are two chain sequences, one propagated by reactions (3) and (4) to give the isomer I, and the other propagated by reactions

(2), (3) and (5) leading ultimately by reaction (6) to the disiletane II.

$$Me_3SiSiMe_3 \longrightarrow 2Me_3Si\cdot \qquad \ldots (1)$$
$$Me_3Si\cdot + Me_3SiSiMe_3 \longrightarrow Me_3SiH + Me_3SiSi(Me_2)\dot{C}H_2 \qquad \ldots (2)$$
$$Me_3SiSi(Me_2)\dot{C}H_2 \longrightarrow Me_3SiCH_2\dot{S}iMe_2 \qquad \ldots (3)$$
$$Me_3SiCH_2\dot{S}iMe_2 + Me_3SiSiMe_3 \longrightarrow Me_3SiCH_2Si(Me_2)H + Me_3SiSi(Me_2)\dot{C}H_2 \qquad \ldots (4)$$
$$Me_3SiCH_2\dot{S}iMe_2 \longrightarrow Me_3Si\cdot + Me_2Si{=}CH_2 \qquad \ldots (5)$$
$$2Me_2Si{=}CH_2 \longrightarrow Me_2Si\langle\ \rangle SiMe_2 \qquad \ldots (6)$$
$$2Me_3SiCH_2\dot{S}iMe_2 \longrightarrow (Me_3SiCH_2SiMe_2)_2 \qquad \ldots (7)$$

[Plus other minor radical-radical and silene-radical addition reactions.]

Scheme 1 — Main reactions in the pyrolysis of hexamethyldisilane.

The pressure-dependence of the product composition is accounted for by the competition between the bimolecular reaction (4) and the unimolecular dissociation (5) of the radical $Me_3SiCH_2SiMe_2\cdot$. The latter reaction, forming a silene intermediate, has a substantially higher activation energy than equivalent olefin-forming reactions of alkyl radicals, giving a short chain length for formation of the 1,3-disiletane (in interesting contrast to carbon chemistry, such short chain reactions are not uncommon in those pyrolyses of organosilicon compounds which are initiated by radical-forming homolyses, thus making accessible non-chain conditions for the measurement of the rates and energetics of initial dissociations of organosilicon molecules [5]). We found that, with an initial pressure of Me_6Si_2 of 0.2 torr, [I]/[II] was 0.4 at 780 K, increasing with decreasing temperature (because of the relatively high value of E_5) to 4.7 at 673 K. With 5 torr of Me_6Si_2, [I]/[II] was 15.5 at 748 K, rising to 57.5 at 621 K. At the substantially higher pressures of sealed-tube experiments, formation of I would be expected to be predominant, as was indeed found [1,6].

Another important difference between silicon and carbon chemistry is that silylenes, the silicon analogues of carbenes, may be produced under quite mild thermolysis conditions from suitable precursors. Silylene formation is minor (but not negligible [3]) in the pyrolysis of $Me_3SiSiMe_3$, but is the major primary reaction in the pyrolysis of other disilanes with a silicon-hydrogen or silicon-halogen bond [7]. Silylene formation is also important in the pyrolysis of monosilanes with silicon-hydrogen bonds. Relevant examples are in Table 1 (much of the current kinetic knowledge on silylene-forming reactions comes from the work of Ring, O'Neal and co-workers [8]).

Besides organosilyl radicals and silylenes, silenes are also important intermediates in organosilicon chemistry. They may be produced by decomposition of large radicals, as in reaction (5), but a more direct route is by thermolysis of siletanes. It is well established [12] that 1,1-dimethyl-

Table 1 — Arrhenius parameters for primary reactions in the pyrolysis of methyl disilanes and monosilanes.

REACTION		log A*	E/kJ mol⁻¹	k_{900K}/s⁻¹	Ref.
Me₃SiSiMe₂H $\longrightarrow$ Me₂S̈i + Me₃SiH	(8)	12.93±0.31	198±3.9	2.70×10¹	9
Me₃SiSiMe₃ $\longrightarrow$ Me₂S̈i + Me₄Si	(9)	13.7±0.7	282±12	2.15×10⁻³	3
Me₃SiSiMe₃ $\longrightarrow$ Me₃Si· + Me₃Si·	(1)	17.2±0.3	337±4	4.37×10⁻³	3
MeSiH₃ $\longrightarrow$ MeSiH + H₂	(10)	15.2	271		8,10,11
MeSiH₃ $\longrightarrow$ S̈iH₂ + CH₄	(11)	14.7	279		8,10
Me₂SiH₂ $\longrightarrow$ Me₂S̈i + H₂	(12)	14.3	285		8,11
Me₂SiH₂ $\longrightarrow$ MeS̈iH + CH₄	(13)	15.0	301		8

*First-order A factors and rate constants are in s⁻¹ and second-order in dm³ mol⁻¹ s⁻¹. For consistency, all Arrhenius parameters are from Ref. 8; concordant parameters are given in the other references cited.

siletane decomposes cleanly to give dimethylsilene (Me$_2$Si=CH$_2$) and ethene:

$$\square\!\!-\!\text{SiMe}_2 \longrightarrow \text{Me}_2\text{Si=CH}_2 + \text{C}_2\text{H}_4$$

But intriguing questions are raised by the thermolysis of siletanes with at least one silicon-hydrogen bond, which show a dramatic increase in mechanistic complexity. In trapping experiments with butadiene, Conlin and co-workers have shown unequivocally that two silylenes, :SiH$_2$ and :Si(Me)H, are present in the pyrolysis of siletane [13], while the major intermediate in the pyrolysis of 1-methylsiletane appeared to be Me$_2$Si: [14]. C$_3$H$_6$ was also found, especially in the pyrolysis of siletane, which was interpreted as evidence for a new primary reaction, direct elimination of a silylene and C$_3$H$_6$, competing with the formation of a silene and C$_2$H$_4$, as in the dimethylsiletane. It was further suggested that silenes with a silicon-hydrogen bond isomerized rapidly and irreversibly to silylenes by a 1,2-hydrogen shift:

$$\text{H(R)Si=CH}_2 \longrightarrow \text{:Si(R)Me} \qquad \ldots\ldots (14)$$

However, it was subsequently realised that conclusions based on yields of butadiene adducts are unreliable, not only because the silene adducts are more unstable thermally than the silylene adducts under these experimental conditions [15,16], but also because rate constants for silylene and silene addition to butadiene are significantly different from each other [16]. This realisation, combined with new experiments in which Me$_2$Si: was generated over a wide range of temperature, in the presence or absence of butadiene, enabled us to conclude [16] that the isomerization reaction (14) was indeed reversible with an equilibrium constant close to unity and an energy barrier of ~170 kJ mol^{-1}, as predicted theoretically [17]. The credit for first suggesting that silylenes may isomerize to silenes belongs to Conlin and Gaspar [18]. We have now confirmed our conclusions by numerical integration [19]. A general mechanism linking the reactions of HMeSi=CH$_2$ (III) and Me$_2$Si: (IV) is given in Scheme 2, with individual Arrhenius parameters in Table 2. Minor reactions in Scheme 2 are denoted by dotted arrows.

In the absence of butadiene, the ultimate products are the disiletanes V and VI, while in the presence of butadiene they are the tetrahydrosilin VII and the silolene VIII. Calculated ratios of [VI]/[V] or [VII]/[VIII] agreed satisfactorily with several independent experiments covering a wide range of conditions, as may be seen from Table 3.

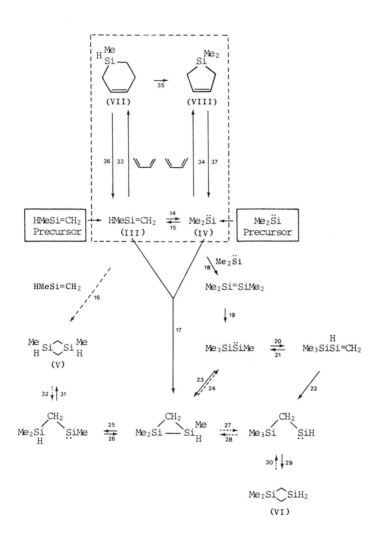

Scheme 2 — The silylene-silene isomerization (with and without trapping).

Table 2 — Estimated Arrhenius parameters for reactions in Scheme 2.

Reaction	14,15	16	17	18	19	20,21	22	23	24	25
log A	13.5	6.6	10	10	13.5	13.5	12.3	12.6	14	13
E/kJ mol	170	0	0	0	180	170	120	165	⩽218	⩽115

Reaction	26	27	28	29,31	30,32	33	34	35	36	37
log A	12.6	14	12.6	12.9	13.5	7	9.5	13.5	14.4	13.5
E/kJ mol	95	⩽199	145	136	255	10	12	252	252	255

Table 3 — Tests of Scheme 2 against experimental results.

Initial Intermediate	No Butadiene Experimental Conditions	Ref.	T/K	[VI]:[V] Observed	Calculated
(III)			760	0.4	0.46
(III)	Stirred flow in 2½	20	850	0.5	0.56
(IV)	atmospheres of nitrogen		760	1.4	1.3
(IV)			850	0.6	0.67
(III)	Flow, 1-5 torr	14	898	0.53	0.56
(III)			925	0.56	0.57
(III)	Nitrogen flow	21	673	0	0.005
(IV)	Vacuum flow	18	873	0.86	0.64
(IV)	<0.1 torr		973	0.6	0.62
(IV)₂	Sealed tube	22	633	3	3
Initial Intermediate	Excess of Butadiene and Conditions	Ref.	T/K	[VII]:[VIII] Observed	Calculated
(IV)	× 10; stirred flow	16	720	<1%	0.0003
(IV)			924	<1%	0.003
(III)			829	~0.8	0.6
(III)	× 8	14	898	~0	0.067
(III)			925	~0	0.014
(III)	N₂ flow	21	673	[VII]≫[VIII]	20
(III)	~1 atmosphere	16	873	[VII]≫[VIII]	12

Closely related to the foregoing is the remarkably specific thermal rearrangement of $(Me_3Si)_2Si$:, which gives the trisilolane IX in 60% yield at 773 K [23]:

Me₃SiSiSiMe₃ —Δ→

CH₂
Me₂Si   SiH₂   (60%)
|
CH₂
SiMe₂
(IX)

We have extended our quantitative model to account for that intriguing result via a trimethylsilyl shift, as shown in the

partial mechanism in Scheme 3 [24]. Arrhenius parameters for
the individual reactions are in Table 4.

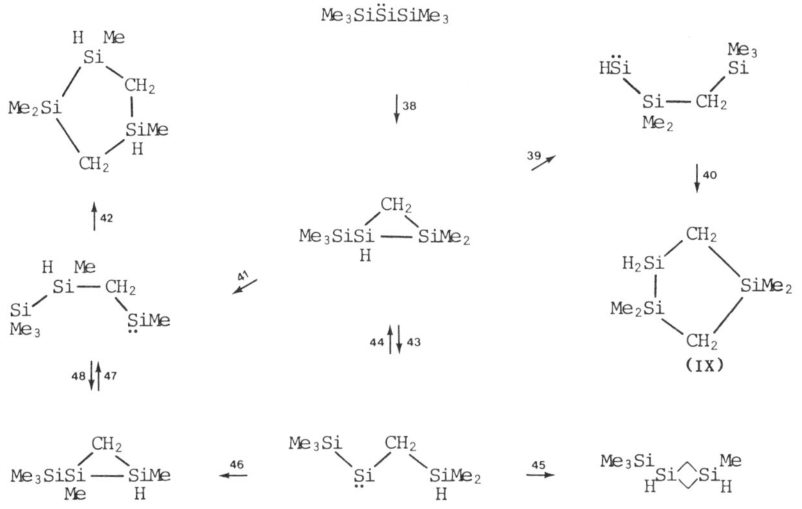

Scheme 3 — Partial mechanism showing route to (IX).

Table 4 — Arrhenius parameters for reactions in Scheme 3.

Reaction	38	39	40	41	42	43	44	45	46	47	48
log A	12.6	13.7	13	13.7	13	13	12.6	12.9	12.7	13	12.6
E/kJ	165	78	90	88	90	24	83	145	136	4	83
Rel. Rate	1.66	1.00	1.00	0.21	0.33	891	891	0.12	0.31	0.50	0.39

We have made some progress in understanding other aspects of
the complex mechanism of pyrolysis of hydridosiletanes through
gas kinetic studies. Our kinetic data for decomposition of
siletane (X) and 1-methylsiletane (XI) and for formation of
ethene and propene from these, are compared with literature
data [12] for 1,1-dimethylsiletane (XII) in Table 5. In the
pyrolysis of the unsubstituted siletane X we also found two
previously unreported products, silane and methylsilane.
Formation of ethene from all three siletanes has about the same
activation energy, suggesting a common primary process followed
by secondary reactions of low activation energy in X and XI.
Formation of propene, having a _lower_ activation energy than
ethene formation, cannot result directly from silicon-carbon
bond cleavage, as previously suggested [13]. An alternative
route invoking known reactions in organosilicon chemistry is
the 1,2 hydrogen-shift with concomitant silicon-carbon bond

Process	Compound	log A (s^{-1})	E/kJ mol^{-1}	k_{800K}/s^{-1}	$k_{relative}$
Formation of C_2H_4	XII	[15.64±·3	262±3	·037]^a	1
	XI	15.8 ±·6	260±8	.067	1.8
	X	16.4 ±·3	264±4	.145	3.9
Formation of C_3H_6	XI	13.4 ±·6	243±8	.004	1
	X	14.4 ±·2	242±3	.040	10
Total decomposition	XII	[15.64	262	.037]^a	1
	XI	~14.4	~233	.143	3.9
	X	~14.0	~220	.431	11.6

^{a} From Ref. 12

Table 5 — Arrhenius parameters and rate constants for pyrolysis of silacyclobutanes.

breaking, first suggested [10] as the route to silylene and
methane in the pyrolysis of methylsilane [reaction (11) in
Table 1]; a better analogy for X or XI is reaction (13) in
Table 1 [8]. Propene would therefore result from secondary
decomposition of a propylsilylene, thus:

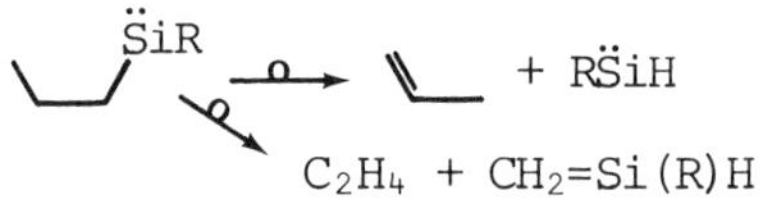

Propylsilylenes are known [8] to decompose to propene, but give
other products as well, thus accounting for the A factors in
Table 5 being lower than A_{13}; release of ring strain would
account for the lower activation energy. Hence, the pyrolysis
of XII is much simpler than that of X or XI because the
dimethylsilene formed from XII cannot isomerize to a silylene,
nor can a propylsilylene be formed. Furthermore, in the
pyrolysis of X or XI, silylenes will insert rapidly into
silicon-hydrogen bonds in the parent molecules, thus initiating
secondary reactions not possible in XII. In the case of X,
with two silicon-hydrogen bonds in the molecule, silylene
chains [8] can then occur, leading to silane and methylsilane,
as observed, and accounting for the increased complexity and
rapidity of that pyrolysis. These mechanistic features of the
pyrolysis of X and XI are summarised in Schemes 4 and 5.
     A further feature is that disiletanes would be expected to
be formed by the reactions exemplified by Scheme 2, rather than
simply by silene dimerization, as in the pyrolysis of XII. The
question of disiletane formation in these pyrolyses is further
complicated by the possibility that hydridodisiletanes may
decompose by similar hydrogen-shift mechanisms to those in
Schemes 4 and 5! In a few preliminary experiments [25] we have
found that the rate constants for decomposition of 1,3-disile-
tane, 1,3-dimethyl-1,3-disiletane, and 1,1,3,3-tetramethyl-1,3-
disiletane were in the ratio of 1863:69:1 at 800 K, which
appears to be in accordance with that suggestion. In the wake

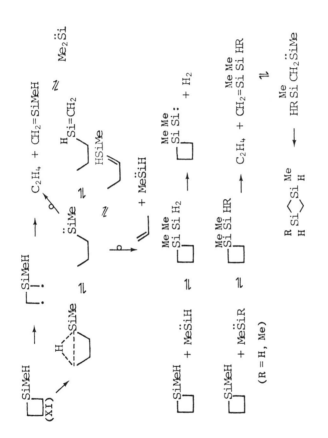

Scheme 4 – Pyrolysis of XI.

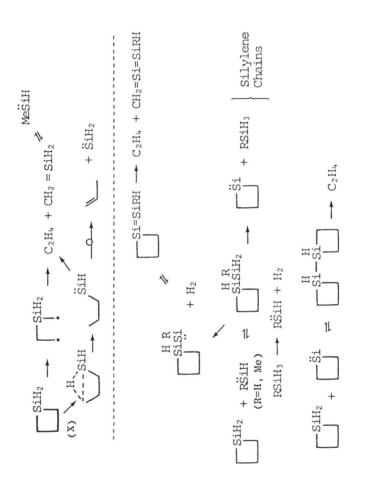

Scheme 5 — Pyrolysis of X.

of the stimulating original experiments by Conlin [13,14], much kinetic and mechanistic work remains to be done in this intriguing area of organosilicon chemistry.

ACKNOWLEDGEMENTS

It is a pleasure to acknowledge the enthusiasm and skill of my co-workers, Dr. Sina Ijadi-Maghsoodi, Andrew Fenton, Kevin J. Hughes and Robert J. Scampton, all of whom were supported by the Science and Engineering Research Council. All of us derived great scientific benefit and stimulus from a three-month visit to our laboratory in 1983 by Professor Thomas J. Barton, in the course of which much of the work described above was done or initiated.

REFERENCES

1. K. Shiina and M. Kumada, J. Org. Chem., 1958, 23, 139.
2. I. M. T. Davidson and A. V. Howard, J. Chem. Soc., Faraday Trans. I, 1975, 71, 69.
3. I. M. T. Davidson, C. Eaborn and J. M. Simmie, J. Chem. Soc., Faraday Trans. I, 1974, 70, 249.
4. I. M. T. Davidson, P. Potzinger and B. Reimann, Ber. Bunsenges Phys. Chem., 1982, 86, 13.
5. A. C. Baldwin, I. M. T. Davidson and M. D. Reed, J. Chem. Soc., Faraday Trans. I, 1978, 74, 2171;
    I. M. T. Davidson and C. E. Dean, unpublished work.
6. H. Sakurai, A. Hosomi and M. Kumada, Chem. Comm., 1968, 930.
7. I. M. T. Davidson, J. Organometallic Chem., 1970, 24, 97.
8. M. A. Ring, H. E. O'Neal, S. F. Rickborn and B. A. Sawrey, Organometallics, 1983, 2, 1891.
9. I. M. T. Davidson and J. I. Matthews, J. Chem. Soc., Faraday Trans. I, 1976, 72, 1403.
10. I. M. T. Davidson and M. A. Ring, J. Chem. Soc., Faraday Trans. I, 1980, 76, 1520.
11. P. S. Neudorfl and O. P. Strausz, J. Phys. Chem., 1978, 82, 241.
12. M. C. Flowers and L. E. Gusel'nikov, J. Chem. Soc. (B), 1968, 419 and 1396.
13. R. T. Conlin and R. S. Gill, J.A.C.S., 1983, 105, 618.
14. R. T. Conlin and D. L. Wood, J.A.C.S., 1981, 103, 1843.
15. R. T. Conlin, at XVII Organosilicon Symposium, Fargo, N. Dakota, 1983.
16. I. M. T. Davidson, S. Ijadi-Maghsoodi, T. J. Barton and N. Tillman, J. Chem. Soc., Chem. Comm., 1984, 478.
17. H. F. Schaefer III, Acc. Chem. Res., 1982, 15, 283.
18. R. T. Conlin and P. P. Gaspar, J.A.C.S., 1976, 98, 868.
19. I. M. T. Davidson and R. J. Scampton, J. Organometallic Chem., 1984, in press.

20. M. J. Blades, I. M. T. Davidson, K. J. Hughes, S. Ijadi-Maghsoodi and R. J. Scampton, to be published.
21. T. J. Barton, S. A. Burns and G. T. Burns, Organometallics, 1982, $\underline{1}$, 210.
22. D. N. Roark and G. J. D. Peddle, J.A.C.S., 1972, $\underline{94}$, 5837.
23. Y. S. Chen, B. H. Cohen and P. P. Gaspar, J. Organometallic Chem., 1980, $\underline{195}$, C1.
24. I. M. T. Davidson, K. J. Hughes and R. J. Scampton, J. Organometallic Chem., 1984, in press.
25. I. M. T. Davidson and S. Ijadi-Maghsoodi, unpublished work.

CHAPTER 7

MECHANISTIC STUDIES OF SILICON ATOMS AND SILYLENES

Peter P. Gaspar Department of Chemistry, Washington University, Saint Louis, Missouri 63130, USA

INTRODUCTION

Reactions of atoms and molecules with low-lying excited states are of considerable interest because they can reveal the consequences of changes in electronic structure at nearly constant energy. The well-known differences between the reactions of singlet and triplet carbenes exemplify the dramatic effects of this 'electronic isomerism'[1]. Silicon atoms and silylenes are of particular importance since comparison of their chemistry with that of carbon atoms and carbenes can help to define the limits of validity of the fundamental mechanistic ideas of organic chemistry, ideas that are couched in universal-sounding terms of symmetry, energy, electron density, orbital shapes, and overlap.

SILICON ATOMS AND THEIR REACTIONS

Study of the gas-phase reactions of high energy silicon atoms generated by the $^{31}P(n,p)^{31}Si$ nuclear transformation has revealed a host of addition, insertion and abstraction reactions, many of which show a strong dependence on kinetic energy and, apparently, on the electronic state of the reacting atoms[2,3].

Silicon atom reactions have proved to be convenient sources of silylenes, many of whose reactions were first found in atomic silicon experiments. Early evidence for the insertion of SiH_2 into Si-H bonds came from the study of the products from recoiling silicon atoms in mixtures of phosphine and silane[4].

$$^{31}Si \xrightarrow[PH_3, SiH_4]{} {}^{31}SiH_2 \xrightarrow{SiH_4} {}^{31}SiH_3SiH_3$$

The acquisition of hydrogens is a high energy process[5], but it has recently been shown that the resulting $^{31}SiH_2$ molecules are in their ground singlet electronic state and at ambient temperature[6]. This was accomplished by comparison of the relative reactivity and its temperature dependence of nucleogenic silylene with that of thermally generated singlet SiH_2.

At present the recoil technique is the only available means for studying the reactions of $^{31}SiH_2$ at room temperature. The importance of being able to work at ambient temperatures is underscored by the study of SiF_2. The characteristic addition reactions of difluorosilylene were long obscured by the secondary transformations that occur under the high temperature conditions at which SiF_2 is generated by the method of Margrave and Timms[7]. It was Tang and coworkers who discovered the addition of difluorosilylene to butadiene, employing stepwise fluorine acquisition by recoiling silicon atoms to generate $^{31}SiF_2$[8-12].

The addition of SiH_2 to butadiene yielding 1-silacyclopent-3-ene was first found in a silicon atom recoil reaction[13,14] and then carried out in a chemical process[14], although thermally generated Me_2Si had earlier been shown to form a silacyclopentene on addition to 2,3-dimethylbutadiene[15].

Addition of silicon atoms to butadiene gives rise to an unusual product and a mechanistic puzzle. It was suggested in 1974 that the major product was silole, a molecule whose synthesis had been attempted by several workers without success over a fifty-year period[14]. Tang and coworkers provided further evidence for its identification by reducing the unknown recoil reaction product to silacyclopentene[16]. A mechanism for silole formation including as a key step the rearrangement of 1-silacyclopent-3-enylidene is based on the observation that 5-silaspiro[4,4]nona-2,7-diene is also formed. An initial 1,2-addition is suggested on the basis of our studies of the silylene-butadiene addition mechanism discussed below.

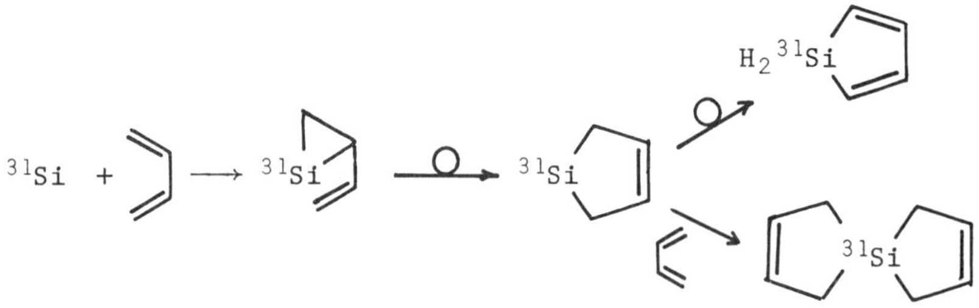

When the silacyclopentenylidene is generated thermally by pyrolysis of 1-methoxy-1-trimethylsilyl-1-silacyclopent-3-ene, the product-forming steps depicted above can be reproduced and macroscopic quantities of products isolated. When this pyrolysis is carried out in liquid butadiene the spiro-adduct is

formed in 32% yield[17], while gas-phase pyrolysis yields silole
which was isolated as its Diels-Alder dimer[18].

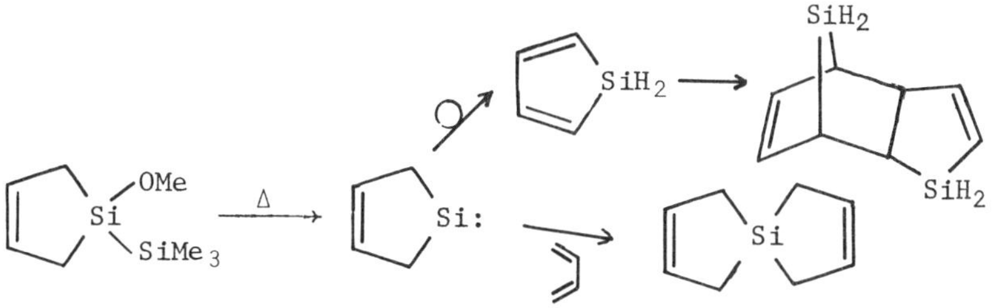

The silacyclopentenylidene-to-silole rearrangement is of
mechanistic interest because it may be an example of a silylene-
to silene interconversion, a process that remains controversial
[19-21] although it was suggested as early as 1974[22]. A high

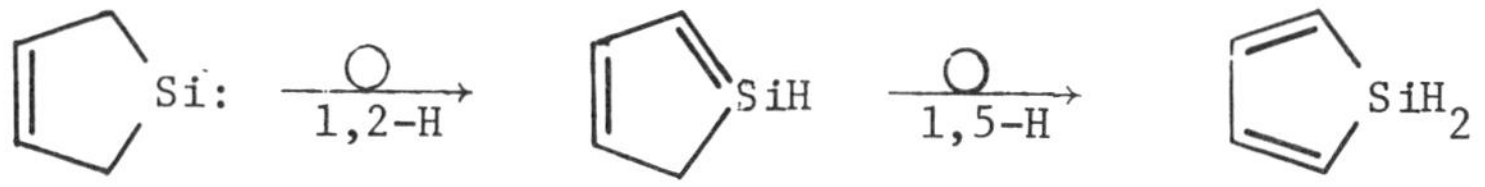

activation barrier has been calculated for this process, but
several examples of silene-to-silylene rearrangements have been
reported[23].

SILYLENE ADDITION TO SUBSTITUTED BUTADIENES

Methyl-substituted butadienes have been useful substrates
for study of the silylene addition mechanism. When there are no
substituents on the terminal carbons, addition of dimethylsilylene leads to high yields of 1,1-dimethyl-1-silacyclopent-3-enes
[24].

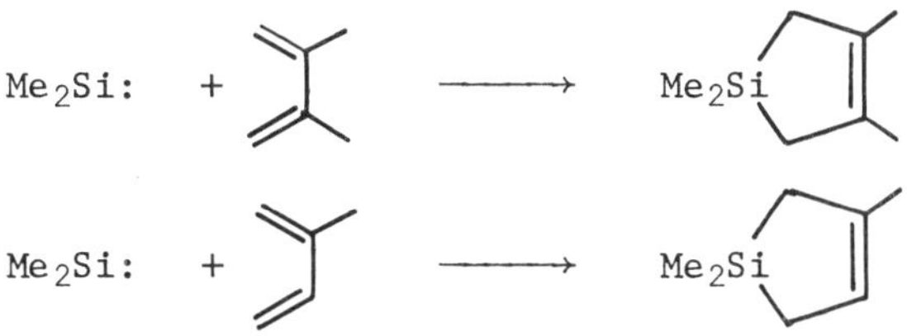

When both terminal carbons carry methyl substituents, only
low yields of 1-silacyclopent-3-enes are formed, and the product
ratio depends strongly on the geometry of the starting diene.
Several points should be noted in reference to the product distribution shown below: 1. Not only are the yields of 1-silacyc-
lopent-3-enes low, but addition is not stereospecific. Hence
concerted 1,4-addition is precluded, and the mechanism for formation of the 1-silacyclopent-3-enes must permit rotation about
the original pi-bonds. 2. The intervention of vinylsilacyclopropane intermediates is strongly indicated by the finding of
products whose structures are most easily rationalized in terms
of carbon-carbon bond cleavage of such species. 3. Concerted

Table I — Product yields (% absolute) from addition of Me_2Si to 2,4-hexadienes.

Me_2Si (structure)	+ Me_2Si (structure)	+ Me_2Si (structure)	+ Me_2Si (structure)	+ Me_2Si Et (structure)
4.2	1.1	10.4	5.3	6.3
1.3	1.4	9.5	11.7	19.8
2		8	3	27.5

1,2-addition of Me_2Si to these terminally substituted butadienes to form vinylsilacyclopropane intermediates is suggested
by the variation in product ratios with a change in geometry of
the diene substrate and by the predominance of a product from
the cis,cis-2,4-hexadiene that is neatly accounted for by a
Cope rearrangement of a cis-substituted silacyclopropane:

4. Cleavage of both carbon-carbon and carbon-silicon bonds of
the silacyclopropane ring leading to the formation of diradical
intermediates can be inferred from the non-stereospecific formation of the 1-silacyclopent-2-ene and 1-silacyclopent-3-ene
products. These can be viewed as resulting from intramolecular
recombination of the radical centers, a process earlier suggested by Sakurai[26]. These considerations lead to the following
proposed mechanism, that is believed to be valid for the addition of various silylenes to a range of substituted 1,3-dienes.
Some small contribution from concerted 1,4-cycloaddition may
occur, but there is a remarkable difference between the addition of dimethylsilylene to 1,3-dienes and the addition of di-
methylgermylene. The addition of Me_2Ge: to trans,trans- and
cis,trans-2,4-hexadienes has been found to be greater than 96%
stereospecific, and thus may be a concerted 1,4-addition[27,28].

The cause of this apparent mechanistic dichotomy is an intriguing problem.

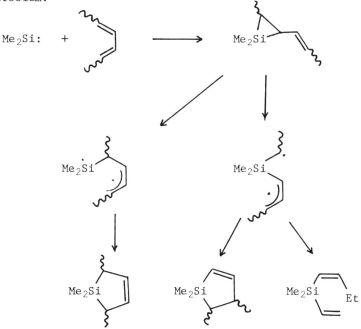

REARRANGEMENTS OF α-SILYLSILYLENES

Rearrangements of silylenes have already been mentioned above, but a particular class of silylene rearrangements, those of α-silylsilylenes, has been the object of close attention in our laboratories for several years[29]. Such reactions can lead to dramatic reshuffling of molecular structure and still be quite specific:

$$(Me_3Si)_2Si: \xrightarrow{\quad\bigcirc\quad} \begin{array}{c} Me_2Si \overset{\displaystyle CH_2}{\underset{\displaystyle H_2C}{|}} \diagdown SiH_2 \\ \diagup SiMe_2 \end{array}$$

60%

Trimethylsilylsilylene came under scrutiny because it could arise from insertion of a recoiling silicon atom into the Si-H bond of trimethylsilane[30]. Reactions of chemically generated $Me_3Si\text{-}\ddot{S}i\text{-}H$ were examined when it was found that the recoil reaction gave unexpected products.

$$Si \; + \; HSiMe_3 \xrightarrow{\text{insertion}} H\text{-}\ddot{S}i\text{-}SiMe_3 \xleftarrow[-HSiMe_3]{\Delta} (Me_3Si)_2SiH_2$$

When $Me_3Si-\ddot{S}i-H$ is generated in the gas-phase at low pressure, the sole product is 1-methyl-1,3-disilacyclobutane, as predicted by a mechanism, first suggested by Wulff, Goure and Barton[31], for a rearrangement involving the formation and decomposition of a disilacyclopropane intermediate[30]. Generation of $Me_3Si-Si-H$ in the presence of several torr of Me_3SiH leads to a decrease in the yield of disilacyclobutane (a product not found from the reactions of recoiling silicon atoms and trimethylsilane) and the formation of products of intermolecular trapping of β-silylsilylenes formed by the rearrangements depictted below. The intermolecular trapping products are found in recoil reaction systems together with $(Me_3Si)_2{}^{31}SiH_2$, the result of two consecutive insertions of ^{31}Si atoms into two Me_3SiH molecules without intervening rearrangements.

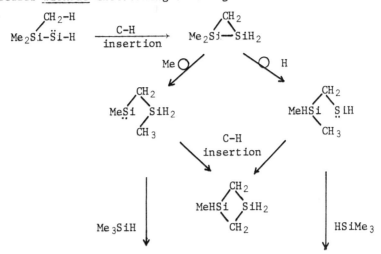

The trapping of $Me-\ddot{S}i-CH_2SiH_2Me$ could be considered as conclusive evidence for the occurence of methyl-migration in the decomposition of a disilirane intermediate, had not a new silylene rearrangement, a transposition, been encountered that competes with the formation of the disilirane intermediate[32,33]. The transposition consists of an interchange of a hydrogen and a methyl group that is mediated by a methyl-shift that interconverts trimethylsilylsilylene and trimethyldisilene. This suggests strongly that the tetramethyldisilene-to-methyl(trimethylsilyl)silylene rearrangement discovered by Barton and coworkers [31] is reversible. The newly discovered transposition is completed by a disilene-to-silylsilylene rearrangement via a hydrogen shift.

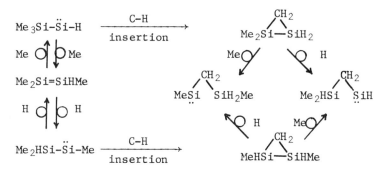

All three of these rearranged silylenes have been trapped by insertion into the Si-H bonds of $HSiMe_3$ and $HSiMe_2Et$ and by addition to butadiene. The disilene intermediate in the transposition has also been trapped by addition to butadiene:

$$Me_2Si=MeHSi \quad + \quad \diagup\diagdown \quad \longrightarrow \quad Me_2Si\text{-}MeHSi\diagup$$

The question concerning the shift of a methyl group in the ring-opening of the 1,1-dimethyl-1,2-disilacyclopropane is due to the recognition that the β-silylsilylene resulting from this process could also arise from a hydrogen-shift in the ring-opening of an isomeric disilirane formed from intramolecular C-H insertion in the transposed α-silylsilylene:

The extent to which each of these two routes contributes to the formation of $Me\text{-}\ddot{S}i\text{-}CH_2SiH_2Me$ has yet to be determined. In the case of the first silylene-to-silylene rearrangement reported by Wulff, Goure and Barton, transposition of $Me_3Si\text{-}\ddot{S}i\text{-}Me$ is a degenerate rearrangement. There is nevertheless an alternative route to the β-silylsilylene that can be formulated as arising via a methyl-shift in a disilirane ring-opening. A silylene-to-silene rearrangement via a hydrogen-shift followed by the shift of a trimethylsilyl group can produce the same β-silylsilylene [34].

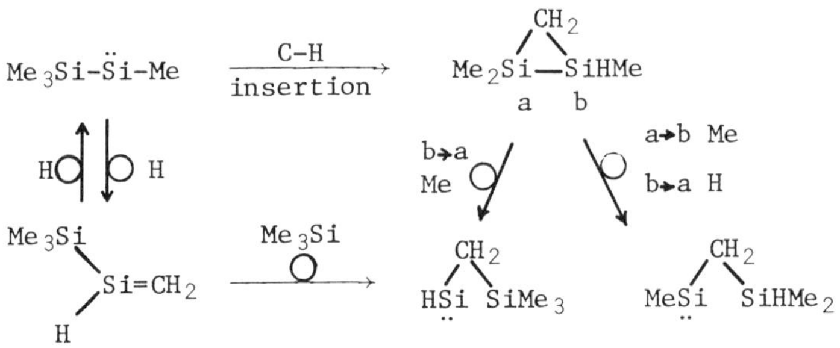

$$Me_3Si\text{-}\overset{..}{Si}\text{-}Me \quad \xrightarrow[\text{insertion}]{C\text{-}H} \quad Me_2\overset{\displaystyle CH_2}{\underset{a}{Si}}\underset{b}{\diagup\!\diagdown}SiHMe$$

$$H\!\bigcirc\,\bigcirc\!H \qquad\qquad \overset{b\to a}{\underset{Me}{\diagup}}\bigcirc \qquad \bigcirc\diagdown\overset{a\to b\ Me}{\underset{b\to a\ H}{}}$$

$$Me_3\overset{\displaystyle Si}{\diagdown}_{Si=CH_2} \xrightarrow{\quad Me_3Si\bigcirc \quad} \overset{CH_2}{H\overset{..}{S}i\diagup\!\diagdown SiMe_3} \qquad \overset{CH_2}{Me\overset{..}{S}i\diagup\!\diagdown SiHMe_2}$$
H

LASER STUDIES OF SILYLENE REACTIONS

The high reactivity and unusual reactions of silylenes have sustained a widespread interest in the elucidation of their reaction mechanisms, but the short lifetimes of most silylenes have made it rather difficult to detect them directly and to properly fix their detailed reaction pathways. We are however entering a new stage in the investigation of silylenes, as well as other reactive intermediates.

The pioneering work of Kumada and coworkers has provided photochemical routes to the generation of silylenes[35], and

$$(Me_2Si)_6 \xrightarrow{h\nu} Me_2Si: + (Me_2Si)_5$$

$$(Me_3Si)_2SiMePh \xrightarrow{h\nu} :SiMePh + Me_3SiSiMe_3$$

the matrix-isolation experiments of Michl, West and coworkers have demonstrated the feasibility of detecting silylenes by electronic spectroscopy[36,37].

The groundwork has therefore been laid for the generation of silylenes by laser photolysis and the monitoring of their concentrations by electronic absorption and emission spectroscopy, thus making possible real-time direct kinetic studies on the nanosecond and, where required, picosecond time scales. A tremendous growth in our knowledge and understanding can be anticipated when the photophysics and photochemistry of silylene generation and the mechanisms of silylene insertion, addition, abstraction and rearrangement reactions will be examined with these powerful techniques.

A start has been made, and two transient absorption spectra are shown below. The transient with an absorption maximum at 440 nm from photolysis of 2-phenylheptamethyltrisilane (Figure 1) is believed to be methyl(phenyl)silylene :SiMePh [38], and one of the two transients from photolysis of dodecamethylcyclohexasilane (Figure 2) is likely to be dimethylsilylene Me_2Si: [36,39,40, 41].

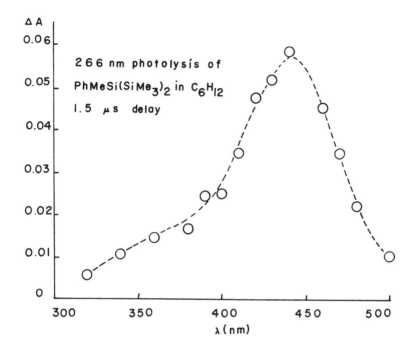

Figure 1

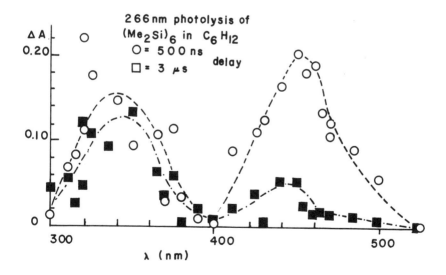

Figure 2

From variation of pseudo-first-order rate constants with substrate concentration, second-order rate constants have been measured for the disappearance of the 440 nm transient formed in the photolysis of $PhMeSi(SiMe_3)_2$, and these are given in Table II. If our identification of the carrier as $:SiMePh$ is correct, these are among the first absolute rate constants for silylene reactions determined at room temperature.

Table II — Second-order rate constants from quenching studies in the 266 nm photolysis of 2-phenylheptamethyltrisilane.

Quencher	Concentration (M)	k $(M^{-1}s^{-1})$ (440 nm)
O_2	1 and 5 x 10^{-3}	3 x 10^8
EtOH	4 to 10 x 10^{-2}	5 x 10^6 to 3 x 10^7
$ClCH_2CH_2Cl$	0.6 to 1.2	2 x 10^6
⋈	5 to 15 x 10^{-2}	1 x 10^5
$EtMe_2SiH$	0.3 to 0.8	1 x 10^5
Et_3SiH	0.6 to 1.5	5 x 10^4

More work is needed before the identification of silylenes via their transient spectra is placed on a firm basis, and this must occur before such kinetic studies can be accepted as reliable sources of mechanistic information.

CONCLUSION

The vigorous efforts of many groups over the past decade have given us a reasonably complete qualitative picture of silylene chemistry. While surprises may be expected, we have a fair idea of what reactions do, or are likely to, occur. We are, nevertheless, only at the stage that carbene chemistry found itself in 1960. We are only beginning to acquire detailed knowledge of transition state structures, electronic and steric substituent effects, and absolute reaction rates are just beginning to be measured. This second stage in the investigation of a class of reactive intermediates is vital, not just for the understanding of silylene reactions but for their control and fullest utilization as synthetic reagents. For the physical organic chemist silylenes possess the important virtue of being slightly less reactive and more selective than carbenes, while closely resembling carbenes in their chemistry. This enhances the probability that kinetic studies can define transition state structures with desired accuracy.

ACKNOWLEDGEMENTS

Silylene chemists are friendly and mutually supportive. Thus I owe many colleagues a debt of gratitude for advice and helpful discussions, but especially Tom Barton and Rob Conlin. The encouragement and financial support of our studies of silicon atom and silylene reactions by the United States Department of Energy is gratefully acknowledged. This is technical report COO-1713-133.

REFERENCES

1. Gaspar, P.P.; Hammond, G.S. "Spin States in Carbene Chemistry," Carbenes, vol. II, R.A. Moss and M. Jones, Jr., eds. Wiley, N.Y., 1975, p. 207.
2. Gaspar, P.P. "Recoil Chemistry and Mechanistic Studies with Polyvalent Atoms," Recent Developments in Biological and Chemical Research with Short-Lived Isotopes, K. Krohn and J. W. Root, eds, ACS, Washington, 1981, p. 3.
3. Gaspar, P.P.; Root, J.W. Radiochimica Acta 1981, 28, 191.
4. Gaspar, P.P.; Pate, B.D.; Eckelman, W.C. J. Am. Chem. Soc. 1966, 88, 3878.
5. Gaspar, P.P.; Markusch, P.; Holten, J.D.; Frost, J.J. J. Phys. Chem. 1972, 76, 1352.
6. Gaspar, P.P.; Konieczny, S.; Mo, S.H. J. Am. Chem. Soc. 1984, 106, 424.
7. Gaspar, P.P.; Herold, B.J. "Silicon, Germanium and Tin Structural Analogs of Carbenes," Carbene Chemistry, 2nd ed., W. Kirmse, ed., Academic, N.Y., 1971.
8. Zeck, O.F.; Su, Y.Y.; Tang, Y.-N. J. Am. Chem. Soc. 1976, 98, 3474.
9. Ferrieri, R.A.; Siefert, E.E.; Griffin, M.J.; Zeck, O.F.; Tang, Y.-N. J. Chem. Soc., Chem. Commun. 1977, 6.
10. Siefert, E.E.; Ferrieri, R.A.; Zeck, O.F.; Tang, Y.-N. Inorg. Chem. 1978, 17, 2802.
11. Gaspar, P.P. Reactive Intermediates (Wiley), 1981, 2, 335.
12. Gaspar, P.P. Ibid, 1984, 3, in press.
13. Gennaro, G.P.; Su, Y.Y.; Zeck. O.F.; Daniel, S.H.; Tang, Y. N. J. Chem. Soc., Chem. Commun. 1973, 637.
14. Gaspar, P.P.; Hwang, R.J.; Eckelman, W.C. Ibid, 1974, 242.
15. Atwell, W.H.; Weyenberg, D.R. J. Am. Chem. Soc. 1968, 90, 3438.
16. Siefert, E.E.; Loh, K.-L.; Ferrieri, R.A.; Tang, Y.-N. J. Am. Chem. Soc. 1980, 102, 2285.
17. Barton, T.J.; Burns, S.A.; Gaspar, P.P.; Chen, Y.-S. Synth. React. Inorg. Met.-Org. Chem. 1983, 13, 881.
18. Chen, Y.-S.; Chari, S.; Boo, B.H. work to be published.
19. Schaefer, H.F. III, Accts. Chem. Res. 1982, 15, 283.
20. Walsh, R., J. Chem. Soc., Chem. Commun. 1982, 1415.
21. Nagase, S.; Kudo, T. Ibid, 1984, 141.

22. Conlin, R.T.; Gaspar, P.P. J. Am. Chem. Soc. 1976, 98, 868.
23. Conlin, R.T.; Kwak, Y.-W. Organometallics, 1984, 3, 918.
24. Lei, D.; Gaspar, P.P. J. Organometal. Chem. 1984, in press.
25. Lei, D.; Gaspar, P.P. Organometallics, submitted for publication.
26. Sakurai, H.; Kobayashi, Y.; Sato, R.; Nakadaira, Y. Chem. Lett. 1983, 1197.
27. Ma, E.C.-1.; Kobayashi, K.; Barzilai, M.; Gaspar, P.P. J. Organometal. Chem. 1982, 224, C13.
28. Schriewer, M; Neumann, W.P. J. Am. Chem. Soc. 1983, 105, 897.
29. Chen. Y.-S.; Cohen, B.H.; Gaspar, P.P. J. Organometal. Chem. 1980, 195, C1.
30. Mo, S.-H.; Holten, J.D.; Konieczny, S.; Ma, E.C.-1.; Gaspar, P.P. J. Am. Chem. Soc., 1982, 104, 1424.
31. Wulff, W.D.; Goure, W.F.; Barton, T.J. Ibid, 1978, 100, 6236.
32. Boo, B.H.; Gaspar, P.P. "Transposition - A New Rearrangement of α-Silylsilylenes via Disilene Intermediates," 187th American Chemical Society National Meeting, St. Louis, Missouri, April 8-13, 1984, ORGN 227.
33. Boo, B.H.; Gaspar, P.P. Organometallics, submitted for publication.
34. Boo, B.H.; Gaspar, P.P.; Ghosh, A.K.; Holten, J.D.; Kirmaier, C.R.; Konieczny,S. "Transposition, A New Silylene Rearrangement, and the Laser Photolysis of Polysilanes," XVII Organosilicon Symposium, North Dakota State University, Fargo, North Dakota, June 3-4, 1983, abstracts, p. 12.
35. Ishikawa, M.; Kumada, M. Revs. Si, Ge, Sn and Pb Cpds. 1979, IV, 7.
36. Drahnak, T.J.; Michl, J.; West, R.; J. Am. Chem. Soc. 1979, 101, 5427.
37. West, R.; Fink, M.J.; Michl, J. Science 1981, 214, 1343.
38. Gaspar, P.P.; Boo, B.H.; Chari, S.; Ghosh, A.K.; Holten, D.; Kirmaier, C.; Konieczny, S. Chem. Phys. Lett. 1984, 105, 153.
39. Chari, S.; Kim, D.H.; Tait, C.D. unpublished work.
40. Nazran, A.S.; Hawari, J.A.; Griller, D.; Alnaimi, I.S.; Weber, W.P. work to be published (private communications from Professor Weber and Dr. Griller).
41. Gaspar, P.P.; Boo, B.H.; Chari, S.; Holten, D.; Kim, D.H.; Lei, D.; Tait, D. "Recent Results in the Chemistry of Silylenes," XVIII Organosilicon Symposium, General Electric Research and Development Center, Schenectady, New York, April 6-7, 1984.

CHAPTER 8

UNSATURATED REACTIVE INTERMEDIATES IN ORGANOSILICON CHEMISTRY – RECENT RESULTS

William P. Weber* and Samih A. Kazoura Department of Chemistry & Loker Hydrocarbon Research Institute, University of Southern California, Los Angeles, CA 90089–1661
Georges Manuel and Guy Bertrand Laboratorie de Organométallique, Université Paul Sabatier,118 route de Narbonne, 31062 Toulouse Cedex, France

INTERMEDIATES

The spectrum of reactive π-bonded silicon intermediates has grown rapidly in the last fifteen year [1]. The following doubly bond species have been generated: silenes [$R_2Si=CR'_2$], [2-4] silanones [$R_2Si=O$], [5-8] silaimines [$R_2Si=NR'$], [9-13] silathiones [$R_2Si=S$], [14-16] disilenes [$R_2Si=SiR_2$], [17-18] and silaphosphimines [$R_2Si=PR'$] [19,20].

Much less work has been reported on diagonal π-bonded silicon transients such as 2-silaallene [$H_2C=Si=CH_2$], [21,22] silaketene [$H_2C=Si=O$], [21] silacarbodiimide [$RN=Si=NR$] [23] and silicon dioxide [$O=Si=O$] [6,21]. Finally, no experimental work has appeared on silicon triply bonded transients such as [$-C\equiv Si-$], [$-Si\equiv Si-$] or [$-Si\equiv N$].

We would like to report our efforts to generate silicon dioxide [$O=Si=O$] and silanitrile [$-Si\equiv N$] intermediates.

SILAACETONITRILE/1,3-SIGMATROPIC REARRANGEMENTS OF SILAIMINE INTERMEDIATES

We undertook a study of the flash vacuum pyrolysis (FVP) of dimethoxymethylsilyl-bis(trimethylsilyl)amine (I) in the

$$
\begin{array}{c}
CH_3O \\
CH_3-Si-N \overset{Si(CH_3)_3}{\underset{Si(CH_3)_3}{\big\langle}} \\
CH_3O
\end{array}
\quad (I)
\longrightarrow (CH_3)_3SiOCH_3 +
\left[
\begin{array}{c}
CH_3 \\
Si=N \\
CH_3O \overset{}{\underset{}{}} Si(CH_3)_3
\end{array}
\right]
$$

$$[CH_3-Si\equiv N] + (CH_3)_3SiOCH_3$$

hope that it would undergo 1,2-elimination of two molecules of trimethylmethoxysilane to yield silaacetonitrile [CH_3-Si≡N]. This approach has been successfully used to generate a number of silenes [24-26].

FVP of I and hexamethylcyclotrisiloxane (D_3) was carried out as 500°C and 10^{-4} mm of pressure. D_3 has been previously shown to efficiently trap silenes, [6,27] silanones, [6,28] silathiones [14,15] and silaimines [11]. In addition to trimethylmethoxysilane, the following major products were isolated from this pyrolysis: 8-methoxy-2,2,4,4,6,6,8-heptamethyl-7-trimethylsilyl-1,3,5-trioxa-7-aza-2,4,6,8-tetrasilacyclooctane (40%) (II) and 2,2,4,4,6,6,8,8-octamethyl-7-dimethylmethoxysilyl-1,3,5-trioxa-7-aza-2,4,6,8-tetrasilacyclooctane (III) (20%). II is the expected product of an insertion reaction of N-trimethylsilyl methylmethoxysilaimine and D_3, while III may arise by reaction of N-dimethylmethoxysilyl dimethylsilaimine with D_3.

N-Dimethylmethoxysilyl dimethylsilaimine probably results from a 1,3-sigmatropic rearrangement of a methyl group from one silicon to the other of the initially formed N-trimethylsilyl methylmethoxysilaimine intermediate [44].

Similar interconversions of silene intermediates via 1,3-sigmatropic rearrangements have been observed [29].

$$\left[[(CH_3)_3Si]_2C=SiPh_2 \right] \longrightarrow \left[\begin{array}{c} CH_3 \\ | \\ Ph_2Si \\ \diagdown \\ (CH_3)_3Si \diagup \end{array} C=Si(CH_3)_2 \right]$$

Apparently the first loss of trimethylmethoxysilane from I
to form N-trimethylsilyl methylmethoxysilaimine is significantly easier than the second loss of trimethylmethoxysilane
to form the desired silaacetonitrile intermediate. Nevertheless, a small amount of 1-aza-2,2,4,4,6,6,8,10,10,12,12,14,14-
tridecamethyl-3,5,7,9,11,13-hexaoxa-2,4,6,8,10,12,14-hepta-
silabicyclo[6,6,0]tetradecane (IV) was found (6%). This
product might result from reaction of silaacetonitrile with
two molecules of D_3.

$$[CH_3-Si\equiv N] \ + \ D_3 \longrightarrow \left[\begin{array}{c} \text{structure} \end{array} \right] + D_3$$

However, there is an alternative mechanism for the formation of IV. An initial 1,2-elimination of trimethylmethoxysilane from II could form 1-aza-2,4,4,6,6,8,8-heptamethyl-
3,5,7-trioxa-2,4,6,8-tetrasilacyclooocta-1-ene, a reactive
silaimine, which could react with D_3 to yield IV. Consistent
with this latter possibility, we have found that co-pyrolysis
of II and D_3 at high temperature results in formation of
trimethylmethoxysilane and IV in 65% yield. On the other hand,
co-pyrolysis of III and D_3 gives only recovered starting
materials. This result is expected since III can not lose
trimethylmethoxysilane in a 1,2-sense.

This work was done in collaboration with Dr. S. A. Kazoura.

A NEW ROUTE TO SILANONES, GENERATION OF SILICON DIOXIDE [O=Si=O]

This work has been carried out jointly with Drs. Georges Manuel and Guy Bertrand, Laboratorie de Organométallique, Université Paul Sabatier, Toulouse, France.

Flash vacuum pyrolysis of 6-oxa-3-sila bicyclo[3,1,0]-hexanes, formed by m-chloroperbenzoic acid (MCPBA) oxidation of 1-silacyclopent-3-enes, leads to 3-silacyclopent-4-en-1-ols, 1,3-dienes and silanone intermediates [30] which undergo insertion reactions into Si-O single bonds of 2,2,4,4-tetramethyl-1-oxa-2,4-disilacyclopentane [31,32] as outlined below.

Initial cleavage of a C-O single bond of the bicyclic epoxide is apparently favored.

By comparison, previous work has shown that 3-silacyclo-[3,1,0]hexanes lead to silacyclohexenes as major products [33]. These products apparently result from initial cleavage of the bridging C_1-C_3 bond to yield a 1,3-diradical which undergoes competitive 1,2- and 1,4-hydrogen migration to yield the observed 1-silacyclohex-3-ene and 1-silacyclohex-2-ene.

Silanone intermediates have previously been generated by reaction of non-enolizable ketones or aldehydes with silenes [5,8,34,35], or disilenes [8,36,37]. These transients have also been generated by reaction of silylenes with sulfoxides [38,39] or epoxides [40].

The fact that the 1-silacyclopent-3-ene starting materials are easily prepared by dissolving metal reduction of 1,3-dienes with magnesium in THF or HMPT in the presence of dichloro-silanes makes this new route to silanones preparatively significant [41,42].

The scope of this new reaction has been demonstrated by the generation and chemical trapping of silicon dioxide [O=Si=O]. Thus FVP of 2,7-dimethyl-2,3,7,8-diepoxy-5-sila-spiro[4,4]nonane [43] yields silicon dioxide which has been trapped by its reaction with two molecules of D_3 to yield 2,2,4,4,6,6,10,10,12,12,14,14-dodecamethyl-2,4,7,8,10,12,14-heptasila-1,3,5,7,9,11,13,15-octaoxaspiro[7,7]pentadecane (V).

(V)

Obviously, an alternative mechanistic possibility is that the reactive silicon-oxygen double bonds are generated in a stepwise manner. At this time we have no evidence in favor or against this alternative possibility.

This precursor to silicon dioxide was prepared in two steps: dissolving metal reduction of isoprene with magnesium metal in THF in the presence of dichlorodiethoxysilane gave 2,7-dimethyl-5-silaspiro[4,4]nona-2,7-diene which was oxidized with MCPBA [43].

Finally, we have found that V reacts further with a second equivalent of [O=Si=O] to yield a most interesting tricyclic siloxane polyhydron: 2,4,6,8,10,12,15,17-octasila-1,3,5,7,9-11,13,14,16,18-decaoxatricyclo[5,5,3,3]octadecane.

While many mechanistic questions remain, from a preparative standpoint, the ease of preparation of the necessary precursor, makes this new reaction an extremely attractive method to generate both silanones and silicon dioxide intermediates. This should facilitate the preparation of new spiro and polycyclic siloxane ring systems.

Acknowledgements: We thank the Air Force Office of Scientific Research, Grant 80-0006, for their support. Dr. Weber and Dr. G. Manuel acknowledge the assitance of a NATO travel grant 845/83.

REFERENCES

1. For a recent review of this area see: Gusel'nikov, L.E.; Nametkin, N.S. Chem. Rev. 1979, 79, 529-577.
2. Nametkin, N.S.; Vdovin, V.M.; Gusel'nikov, L.E.; Zav'yalov, V.I. Izv. Akad. Nauk SSSR Ser. Khim 1966, 584.
3. Gusel'nikov, L.E.; Flowers, M.C. J. Chem. Soc. Chem. Commun. 1967, 64.
4. Flowers, M.C.; Gusel'nikov, L.E. J. Chem. Soc. B, 1968, 419.
5. Roark, D.N.; Sommer, L.H. J. Chem. Soc. Chem. Commun. 1973, 167.
6. Golino, C.M.; Bush, R.D.; Sommer, L.H. J. Am. Chem. Soc. 1975, 97, 7371.
7. Davidson, I.M.T.; Thompson, J.F. J. Chem. Soc. Chem. Commun. 1971, 251.
8. Barton, T.J.; Kilgour, J.A. J. Am. Chem. Soc. 1974, 96, 2278.
9. Elseikh, M.; Sommer, L.H. J. Organometal. Chem. 1980, 186, 301.
10. Parker, D.R.; Sommer, L.H. J. Organometal. Chem. 1976, 110, C1.
11. Parker, D.R.; Sommer, L.H. J. Am. Chem. Soc. 1976, 98, 618.
12. Klingebiel, U. Chem. Ber. 1978, 111, 2735.
13. Kazoura, S.A.; Weber, W.P. J. Organometal. Chem. 1984, in press.
14. Sommer, L.H.; McLick, J. J. Organometal. Chem. 1975, 101, 171.
15. Soysa, H.S.D.; Weber, W.P. J. Organometal. Chem. 1979, 165, C1.
16. Carlson, C.W.; West, R. Organometallics, 1983, 2, 1798.
17. Roark, D.N.; Peddle, G.J.D. J. Am. Chem. Soc. 1972, 94, 5837.
18. Smith, C.L.; Pounds, J. J. Chem. Soc. Chem. Commun. 1975, 910.
19. Couret, C.; Escudie, J.; Satge, J.: Andriamizaka, J.D.; Saint-Roch, B. J. Organometal. Chem. 1979, 182, 9.
20. Cowley, A.H.; Newman, T.H. Organometallics, 1982, 1, 1412.
21. Bertrand, G.; Manuel, G.; Mazerolles, P. Tetrahedron, 1978, 34, 1951.
22. Barthelat, J.C.; Trinquier, G.; Bertrand, G. J. Am. Chem. Soc. 1979, 101, 3785.
23. Ando, W.; Tsumaki, H.; Ikeno, M. J. Chem. Soc. Chem. Commun. 1981, 597.
24. Barton, T.J.; Vuper, M. J. Am. Chem. Soc. 1981, 103, 6788.
25. Barton, T.J.; Burns, G.T.; Arnold, E.V.; Clardy, J. Tetrahedron Lett. 1981, 7.

26. Burns, G.T.; Barton, T.J. J. Organometal. Chem. 1981, 216, C5.
27. Golino, C.M.; Bush, R.D.; On, P.; Sommer, L.H. J. Am. Chem. Soc. 1975, 97, 1957.
28. Okinoshima, H.; Weber, W.P. J. Organometal. Chem. 1978, 149, 279.
29. Eaborn, C.; Happer, D.A.R.; Hitchcock, P.B.; Hopper, S.P.; Safa, K.D.; Washburne, S.S.; Walton, D.R.M. J. Organometal. Chem. 1980, 186, 309.
30. Bertrand, G.; Manuel, G. XVII Organosilicon Symposium, Fargo, North Dakota, June 1983.
31. Okinoshima, H.; Weber, W.P. J. Organometal. Chem. 1978, 144, 165.
32. Lane, T.H.; Frye, C.L. J. Organometal. Chem. 1979, 172, 213.
33. Manuel, G.; Faucher, A.; Mazerolles, P. J. Organometal. Chem. 1984, 264, 127.
34. Valkovich, P.B.; Weber, W.P. J. Organometal. Chem. 1975, 99, 231.
35. Davidson, I.M.T.; Dean, C.E.; Lawrence, F.T.; J.C.S. Chem. Commun. 1981, 52.
36. Barton, T.J.; Kilgour, J.A. J. Am. Chem. Soc. 1976, 98, 7231.
37. Fink, M.J.; DeYoung, D.J.; West, R. J. Am. Chem. Soc. 1983, 105, 1070.
38. Soysa, H.S.D.; Okinoshima, H.; Weber, W.P. J. Organometal. Chem. 1977, 133, C17.
39. Alnaimi, I.S.; Weber, W.P. J. Organometal. Chem. 1983, 241, 171.
40. Barton, T.J.; Goure, W.F. J. Organometal. Chem. 1980, 199, 33.
41. Manuel, G.; Mazerolles, P.; Cauquy, G. Syn. React. Inorg. Metal. Org. Chem. 1974, 4, 133.
42. Weyenberg, D.R.; Toporcer, L.H.; Nelson, L.E. J. Org. Chem. 1968, 33, 1975.
43. Terunuma, D.; Hatta, S.; Araki, T.; Ueki, T.; Okazaki, T.; Suzuki, Y. Bull. Chem. Soc. Japan, 1977, 50, 1545.
44. Clegg, W.; Klingebiel, U.; Sheldrick, G.M. Z. Naturforsch. 37b, 423 (1982).

CHAPTER 9

CHEMISTRY OF SMALL RING POLYSILANES

Hamao Watanabe and Yoichiro Nagai Department of Applied Chemistry, Faculty of Engineering, Gunma University, Kiryu, Gunma 376, Japan

The chemistry of small ring peralkylsilanes is an intriguing subject of current interest because of the unique physical and chemical properties of such compounds. Although a number of publications are available on the chemistry of permethylcyclopolysilanes, studies of peralkylcyclopolysilanes bearing alkyl groups other than methyl are so far limited to several series of rings. This paper deals with the preparation of a series of peralkylcyclopolysilanes from various dialkyldichlorosilanes and tetraalkyldichlorodisilanes and also with some physical and chemical properties of cyclopolysilanes.

I. LITHIUM-MEDIATED REDUCTIVE COUPLINGS OF DIALKYLDICHLORO-SILANES AND TETRAALKYL-1,2-DICHLORODISILANES

Reductive coupling of dichloromonosilanes and 1,2-dichlorodisilanes in an etheral solvent with an alkali metal such as Li is the most simple and convenient method for preparing cyclopolysilanes. Reactions of dichlorosilanes (I) with a 10-20% excess of Li in THF at 0° to room temperature gave the corresponding peralkylcyclopolysilanes $[R^1R^2Si]_n$ (n=3-7) in workable yields (Eq. 1) [1]. Reactions using 1,2-dichlorodisilanes (II) also afforded similar results to those from dichlorosilanes (I) (Eq. 2) [2,3].

$$nR^1R^2SiCl_2 \; + \; 2nLi \; \longrightarrow \; [R^1R^2Si]_n \; + \; 2nLiCl \qquad (1)$$
$$(I)$$

$$n(R^1R^2SiCl)_2 \; + \; 2nLi \; \longrightarrow \; 2[R^1R^2Si]_n \; + \; 2nLiCl \qquad (2)$$
$$(II)$$

Cyclopolysilanes in Table 1 are the most stable rings among their homologs and a comparizon of the steric bulk of the substituents with ring size (n) is given. It has been shown that

a mixture of a series of cyclopolysilanes gives thermodynamically the most stable ring as the major product via rapid equilibrium when excess lithium is employed [1,4,5]. Thus, pentamers
from the reactions of two 1,2-dichlorodisilanes (No.7,8) were
found to be the thermodynamic products. It should be noted that
the reaction of (^{t}BuCH$_2$)$_2$SiCl$_2$ gave the first peralkylcyclotrisilane, [(^{t}BuCH$_2$)$_2$Si]$_3$ (No.16).

Table 1 — Relationship between sum of the Es(Si) values [6] for two substituents on silicon
and ring size in cyclopolysilanes.

| No. | Starting chlorosilane | Substituents | | $\sum$Es(Si) | n |
		R^1	R^2		
1	(I); (II)	Me	Me	0.00	6
2	(I)	Pr	Me	−0.215	6
3	(I)	Et	Et	−0.298	5
4	(I)	^{i}Bu	Me	−0.405	5
5	(I)	Pr	Pr	−0.432	5
6	(I)	Bu	Bu	−0.450	5
7	(II)	^{i}Pr	Me	−0.556	5
8	(II)	Me$_3$SiCH$_2$	Me		5
9	(II)	^{t}BuCH$_2$	Me		4
10	(I)	^{i}Bu	^{i}Bu	−0.810	4
11	(I); (II)	^{i}Pr	^{i}Pr	−1.112	4
12	(I); (II)	^{t}Bu	Me	−1.46	4
13	(I)	^{t}Bu	Pr	−1.676	4
14	(I)	^{s}Bu	^{s}Bu		4
15	(I)	Me$_3$SiCH$_2$	Me$_3$SiCH$_2$		4
16	(I)	^{t}BuCH$_2$	^{t}BuCH$_2$		3
17	(I)	^{t}Bu	^{t}Bu	−2.92	2 (?)

The reaction of dichlorodisilanes having bulky substituents
(No.9,11,12) gave tetramers which might be formed via the dimerization [7] of disilenes generated from lithiochlorodisilanes
by β-elimination of LiCl [8,9], as shown in the example of (^{t}Bu-
MeSiCl)$_2$ (No.12) (Eq. 3):

$$(^tBuMeSiCl)_2 \xrightarrow{\text{Li}} \text{Li}\overset{.}{\text{Si}}\overset{.}{\text{Si}}\text{Cl} \rightarrow \underset{\text{Me}}{\overset{^tBu}{}}\text{Si}=\text{Si}\underset{\text{Me}(^tBu)}{\overset{^tBu(Me)}{}} \rightarrow [^tBuMeSi]_4 \quad (3)$$

(cis;trans)   (Four isomers)

It is worthwhile to note that unsymmetrically substituted
1,2-dichlorodisilanes furnish a new type of ring products (Eq.
4).

$$^tBuMe\overset{.}{\text{Si}}\overset{.}{\text{Si}}Me_2 \xrightarrow{\text{Li}} [(^tBuMeSi)_2(Me_2Si)_3] + [(^tBuMeSi)_3(Me_2Si)_2] \quad (4)$$
ClCl

PHYSICAL AND CHEMICAL PROPERTIES OF SOME CYCLOPOLYSILANES

(1) Structural features by X-ray analyses. In the crystallogra-
phical data for cyclopolysilanes, bond distances and ring shapes
in molecules show the most interesting features. Such data ob-

served so far are shown in Fig. 1, together with our data. It is seen that as a general trend the smaller the ring size the longer the Si-Si bond lenths and all the lengths except for

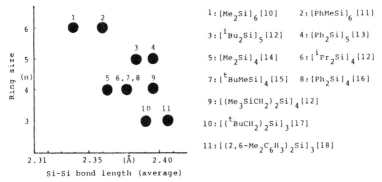

1: $[Me_2Si]_6$ [10]   2: $[PhMeSi]_6$ [11]

3: $[^iBu_2Si]_5$ [12]   4: $[Ph_2Si]_5$ [13]

5: $[Me_2Si]_4$ [14]   6: $[^iPr_2Si]_4$ [12]

7: $[^tBuMeSi]_4$ [15]   8: $[Ph_2Si]_4$ [16]

9: $[(Me_3SiCH_2)_2Si]_4$ [12]

10: $[(^tBuCH_2)_2Si]_3$ [17]

11: $[(2,6-Me_2C_6H_3)_2Si]_3$ [18]

Figure 1 — Relationship between (n) and Si—Si bond length.

$[Me_2Si]_6$ are longer than the normal range of Si-Si single bonds (ca. 2.34 Å). The increase in the length going from Si_6, Si_5, Si_4 to Si_3 is probably the result of increasing steric repulsions between the substituents on silicon and hence of ring strain. Further, for a particular ring size the lengths depend primarily on the steric bulk of the substituents in each series.

Striking features were observed on the ring shapes for the cyclotetrasilanes. The molecular shapes of $[Me_2Si]_4$ [14] and $[(Me_3Si)_2Si]_4$ [19] were reported to be planer, while the other cyclotetrasilanes were shown to be of folded shapes with varing dihedral angles (12-39°). It is natural to consider that usually the most stable structure of these molecules would be the folded one because of the four-valence bond nature in silicon.   The reason why such different structures are possible in the series is not clear at present. However, the molecular structures in crystals would depend upon various controlling factors, such as ring strains, steric repulsions between substituents, conformational requirements of the molecules attributable to the conditions under which crystals were produced (temperature, phase, solvent), etc.

Interestingly, ^1H NMR signals for the methylene hydrogens of $[(^tBuCH_2)_2Si]_3$ at room temperature appear at δ 0.93 ppm and are temperature-dependent. Below -20°, the hydrogens are magnetically non-equivllent and two pairs of AX-type signals were observed at δ 0.44 ppm for H_A and at δ 1.44 ppm for H_X (J_{AX} 15.2 Hz), respectively. The non-equivalence of the hydrogens can be explanined in terms of the hindered rotation around the Si-C bond-(s) by the bulky neopentyl groups.

(2) Correlation of ring size with UV absorption spectra.   The electronic properties of cyclopolysilanes are generally recognized to be associated with the delocalized σ-bonding frameworks in the ring systems and the energy transitions of the bonding electrons can be observed in UV absorptions arising from either

$\sigma(Si-Si)-\sigma^*$ or $\sigma(Si-Si)-\pi^*(d)$ excitations. Thus, it is of great interest to compare the UV absorption spectra of cyclopolysilanes of different ring size. The UV spectra of the cyclopolysilanes listed in Table 2 reveal that the lowest energy transitions for $[R^1R^2Si]_7$, $[R^1R^2Si]_6$, $[R^1R^2Si]_5$, $[R^1R^2Si]_4$ and $[R^1R^2Si]_3$ oc-

Table 2 — UV spectra for peralkylcyclopolysilanes, $[R^1R^2Si]_n$.

Cyclopolysilane			Absorption maxima
R^1	R^2	n	λ_{max} nm(ε)
Me	Me	7	242(5000), 219(10000)
Pr	Pr	"	242(6200)
Me	Me	6	258sh(1300), 233(5800)
Pr	Me	"	257sh(2800), 232(7700)
Me	Me	5	275(700), 264(800)
Et	Et	"	265sh(1400)
Pr	Pr	"	260(1300)
^iBu	^iBu	"	262(1300)
^iBu	^iBu	"	260sh(2300)
^iPr	^iPr	4	290sh(200)
^iBu	^iBu	"	290sh(200)
^sBu	^sBu	"	290sh(200)
^tBu	Me	"	301(190), 262sh(1900), 247sh(2080)
^tBu	Pr	"	294(370), 257sh(2600), 217(169000)
Me_3SiCH_2	Me_3SiCH_2	"	308(260), 263sh(1800), 223(31000)
$^tBuCH_2$	$^tBuCH_2$	3	310sh(330), 265sh(2100), 240sh(7800)

cur in the 240-250, 250-260, 260-280, 290-310 and 300-330 nm regions with varying extinction coefficient, respectively [see also II-(4)]. West et al. previously pointed out that decreasing the ring size in $[R^1R^2Si]_n$ (n=4,5) destabilises the $\sigma(Si-Si)$ bonding levels and shifts the absorptions to longer wavelengths [20]. The bathochromic shifts with decreasing ring size for the cyclopolysilanes described above is in keeping with the trend in the permethylcyclopolysilanes $[Me_2Si]_n$ (n=5-7) [21].

(3) Electrophilic and oxidative cleavage of Si-Si bonds. Since the Si-Si bonds of cyclopolysilanes can be expected to undergo a variety of electrophilic and oxidative cleavage reactions, it is quite significant to investigate such chemical properties. However, there are no reports on the study of air-stable small rings, except for $[^tBuMeSi]_4$ [22]. We investigated some chemical properties of $[^iPr_2Si]_4$ as an example and found that this compound readily undergoes ring-opening reactions at room temperature, as expected, as follows [23]:

$[^iPr_2Si]_4(A) \xrightarrow{MeCOCl/AlCl_3/CCl_4} Cl(^iPr_2Si)_4Cl$ (93%); (A) $\xrightarrow{I_2/PhH}$ $I(^iPr_2Si)_4I$ (95%); (A) $\xrightarrow{HCl/PhH} H(^iPr Si)_4Cl$ (95%); (A) $\xrightarrow{I_2/PhH}$ $\xrightarrow{H_2O} (^iPr_2Si)_4O$ (13%); (A) $\xrightarrow[\text{(Convers. 100\%)}]{m-ClC_6H_4CO_3H} (^iPr_2Si)_4O_n$ (n=1-4)

On the other hand, this compound remained intact after the following treatments: $HCCl_3/130°$; $O_2/PhH/80°$; conc. $H_2SO_4/80°$; $HSiCl_3/H_2PtCl_6/xylene/130°$.

A kinetic study for the reactions of a series of cyclopolysilanes with I_2 has been done by means of the spectrophotometric method using the absorption of I_2 and relative reactivities (k_{rel}, 20°) and activation parameters were determined. The values for the k_{rel} and $\Delta G^{\ddagger}$ listed in Table 3 may reflect the subtle balance of the effects due to the ring strain and steric factors of the substituents. Characteristically, the reaction of $[(^tBuCH_2)_2Si]_3$ with I_2 was so fast that the rate could not be determined by the same way. The ring strain of the three-membered ring account for the exceedingly fast ring-opening reaction. The decreasing trend in Ea values with decreasing ring size could be interpreted in terms of the larger ring strains in the smaller rings. The $\Delta S^{\ddagger}$ values indicate that the contribution of the steric factors to the transition state appears to be fairly large.

Table 3 — Relative reactivities and activation parameters (CH_2Cl_2).

Compound X	$k_x/k_{Me_6Si_2}$ (20°)	$\Sigma Es\,(Si)$	Ea (Kcal/mol)	$\Delta S^{\ddagger}$ (e.u.)	$\Delta G^{\ddagger}(20°)$ (Kcal/mol)
$[^tBuMeSi]_4$	241	-1.46	5.29	-34.6	16.1
$[^iPr_2Si]_4$	237	-1.112	6.56	-38.9	16.1
$[^sBu_2Si]_4$	58		6.00	-39.3	16.9
$[^tBuPrSi]_4$	1	-1.676	7.37	-42.4	19.2
$[Bu_2Si]_5$	11	-0.450	7.49	-37.4	17.9
$[Pr_2Si]_5$	8	-0.432	7.69	-37.2	18.0
$[Me_2Si]_6$	1	0.0	10.0	-33.2	19.1
$[Pr_2Si]_7$	0.04		13.6	-26.8	20.9
$Me_3SiSiMe_3$	1		9.94	-36.8	20.1

One of the most attracting properties of peralkylcyclopolysilanes may be their oxidation potentials, which involves direct removal of one bonding electron from the $\sigma(Si-Si)$ bond. Thus, the first oxidation potentials (Epa) of cyclopolysilanes by

Table 4 — Oxidation potentials by CV method and lowest energy transitions.

Compound	Epa (V) MeCN	CH_2Cl_2	λ_{max} (nm)
$[(^tBuCH_2)_2Si]_3$	+0.44	+0.72	310(sh)
$[^tBuMeSi]_4$	+0.94		300
$[^iPr_2Si]_4$	+1.00	+1.24	290(sh)
$[^iBu_2Si]_4$	+1.02	+1.12	290(sh)
$[^sBu_2Si]_4$	+1.10	+1.23	290(sh)
$[^tBuPrSi]_4$	+1.04		294
$[Bu_2Si]_5$	+1.06	+1.30	262
$[Pr_2Si]_5$	+1.42	+1.45	260
$[Me_2Si]_6$	+1.45	+1.65	258
$[Pr_2Si]_7$	+1.40	+1.40	242

means of cyclic voltammetry (CV method) were determined in MeCN
and CH_2Cl_2. The oxidation potentials are listed in Table 4,
which shows that the trend in the Epa values is in good accord
with that of the activation energies (Table 3). The oxidation
potentials (Epa) reflect the Si-Si bonding levels of these cyclopolysilanes and the ring strain in the smaller rings makes the
levels (HOMO) to push up to higher ones, resulting in lowering
the oxidation potentials. The lowest-oxidation potential for
$[(^tBuCH_2)_2Si]_3$ is in harmony with the result of the fastest rate
in the ring-opening reactions for the series.

(4) Photochemical reactions of cyclotrisilane and cyclotetrasilane derivatives. During the past few years, preparation and
isolation of stable disilene and cyclotrisilane derivatives have
been achieved by a few groups [18, 24-26]. However, all
the disilenes and cyclotrisilane synthesised so far have aryl
substituents on silicon and are not always convenient for as-
sessing the electronic properties of the Si_3 and Si=Si frameworks, because of the perturbation caused by the aryl substituents. As mentioned above, we succeeded to isolate the first peralkylcyclotrisilane $[(^tBuCH_2)_2Si]_3$, which afforded on photolysis
the first peralkyldisilene, $(^tBuCH_2)_2Si=Si(CH_2{}^tBu)_2$ [27].

Thus, upon irradiation (254 nm) at room temperature, a cyclohexane solution of $[(^tBuCH_2)_2Si]_3$ became yellow (λ_{max} 400 nm).
This absorption band gradually dissipated on stopping irradiation
and was observed again on reirradiation. Introduction of air
into the coloured solution resulted in the instant disappearance
of the colour. Irradiation of an enough time gave a mixture containing $(^tBuCH_2)_2SiH_2$, $H[(^tBuCH_2)_2Si]_2H$ and $H[(^tBuCH_2)_2Si]_3H$ as
major products along with small amounts of other products.   In
the next, irradiation of the cyclotrisilane in an EtOH/c-C_6H_{12}
mixture (1:3) gave $H[(^tBuCH_2)_2Si]_2OEt$ and $H[(^tBuCH_2)_2Si]OEt$.
From the above observations, it was concluded that the cyclotri-

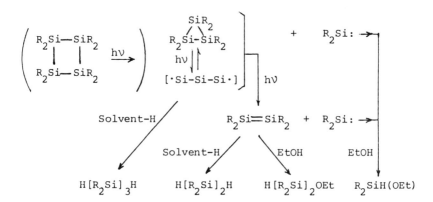

Scheme 1 — Possible scheme for the photolyses of peralkylcyclotrisilane and cyclotetrasilane
derivatives.

silane photochemically decomposed, directly and via trisilane diradical ($\cdot$SiSiSi$\cdot$), into disilene ($^tBuCH_2$)$_2$Si=Si(CH$_2$$^tBu$)$_2$ and silylene [:Si(CH$_2$$^tBu$)$_2$], from which the hydrosilane products were derived as shown in Scheme 1.

Since it has been shown that tetra-aryldisilenes have an intense yellow colour (λ_{max} 420-430 nm) [18,24-26], the absorption maximum exhibited by tetraneopentyldisilene is attributable to the Si=Si frameworks and ca. 20-30 nm shorter than those of the aryldisilenes. It should be emphasised that both the Si$_3$ and Si=Si systems are chromophores per se and the peralkyldisilene was found to be a coloured compound. In connection with this finding, shortly after our publication [27], Masamune et al. reported that compound [(Et$_2$CH)$_2$Si]$_3$ (λ_{max} 328 nm) also undergoes photolysis to furnish the corresponding peralkyldisllene (λ_{max} 390 nm) [28].

Similarly, the photolysis of octaisopropylcyclotetrasilane [$^iPr_2$Si]$_4$ was investigated in cyclohexane and in the presence of ethanol [29]. Thus, the compound was found to undergo ring contraction [30], giving successively [$^iPr_2$Si]$_3$ and $^iPr_2$Si=Si-$^iPr_2$ with extrusion of :Si$^iPr_2$ (Scheme 1). During the photolysis the cyclohexane solution showed a new intense absorption band at λ 320 nm intensity of which rapidly increased and reached the maximum. The colourless solution turned to yellow (λ_{max} 400 nm) and the intensity of the band also gradually increased with increasing reaction time. The absorption bands at λ_{max} 320 nm and λ_{max} 400 nm were attributable to [$^iPr_2$Si]$_3$ and $^iPr_2$Si=Si$^iPr_2$, respectively.

Finally, in order to observe the spectrophotometric properties of the Si$_3$ and Si=Si frameworks, photochemical reactions of peralkylcyclotetrasilanes [R^{1}R^{2}Si]$_4$ having various alkyl groups other than isopropyl group were carried out. Generally, the Si$_3$ and Si=Si frameworks were found to exhibit their absorption bands corresponding to the lowest energy transitions at λ_{max} 300-330 and at λ_{max} 390-400 nm, respectively [31].

The authors are grateful to a number of co-workers who collaborated with them in the experimental work described above. Thanks are also due to Ms. M. Goto of National Chemical Laboratory for Industry for X-ray crystallographic analyses, to Prof. T. Sato of Tokyo Metropolitan University for measurement of CV oxidation potentials and helpful suggestions, and to Prof. R. Okazaki and Dr. T. Kawashima of the University of Tokyo for measurement of low-temperature NMR spectra.

REFERENCES

1. Watanabe, H.; Muraoka, T.; Kageyama, M.; Yoshizumi, K.; Nagai, Y. Organometal., 1984, 3, 141 and the references cited therein.
2. Stolberg, U.G.Z. Angew. Chem., 1963, 75, 206.
3. Watanabe, H.; Inose, J.; Fukushima, K.; Kougo, Y.; Nagai, Y. Chem. Lett., (1983), 1711.
4. Carlson, C.W.; West, R. Organometal., 1983, 2, 1972.

5. Brough, L.F.; West, R. J. Organometal. Chem., 1980, 194, 139.

6. Cartledge, F.K. Organometal., 1983, 2, 425.

7. Matsumoto, H.; Arai, T.; Watanabe, H.; Nagai, Y. J. Chem. Soc., Chem. Sommun., (1984), in press.

8. Weidenbruch, M.; Schäfer, A.; Thom, K.L. Z. Naturforsch., Teil B, 1983, 39, 1695.

9. Watanabe, H. The 48th Meeting of the Chemical Society of Japan, August 30, 1983, Sapporo, Hokkaido; Abstracts II (Symposium) 4Z03, pp. 970-971 and The 30th Symposium on Organometallic Chemistry of the Chemical Society of Japan, November 1-2, 1983, Kyoto; Abstracts A107, pp. 19-21.

10. Carrell, H.L; Donohue, J. Acta Cryst., 1972, B28, 1566.

11. Chen, S.-M.; David, L.D.; Haller, K.J.; Wadsworth, C.L.; West, R. Organometal., 1983, 2, 409.

12. Watanabe, H.; Muraoka, T.; Kougo, Y.; Kato, M.; Kuwabara, H.; Okawa, T.; Nagai, Y.; Goto, M. to be published.

13. Párkányi, L.; Sasvári, K.; Declercq, J.-P.; Germain, G. Acta Cryst., 1978, B34, 3678.

14. Kartky, C.; Schuster, H.G.; Hengge, E. J. Organometal. Chem., 1983, 247, 253.

15. Hurt, C.J.; Calabrese, J.C.; West, R. J. Organometal. Chem., 1975, 91, 273.

16. Párkányi, L.; Sasvári, K.; Barta, T. Acta Cryst., 1978, B34, 883.

17. Watanabe, H.; Kato, M.; Okawa, T.; Nagai, Y.; Goto, M. J. Organometal. Chem., 1984, in press.

18. Masamune, S.; Hanzawa, Y.; Murakami, S.; Bally, T.; Blount, J.F. J. Am. Chem. Soc., 1982, 104, 1150.

19. Chen, Y.-S.; Gaspar, P.P. Organometal., 1982, 1, 1410.

20. Biernbaum, M.; West, R. J. Organometal. Chem., 1977, 131, 179.

21. Carberry, E.; West, R.; Glass, G.E. J. Am. Chem. Soc., 1969, 91, 5446.

22. Biernbaum, M.; West, R. J. Organometal. Chem., 1977, 131, 189.

23. Watanabe, H.; Muraoka, T.; Kageyama, M.; Nagai, Y. J. Organometal. Chem., 1981, 216, C45.

24. West, R.; Fink, M.J.; Michl, J. Science, 1981, 214, 1343.

25. Boudjouk, P.; Han, B.-H.; Anderson, K.R. J. Am. Chem. Soc., 1982, 104, 4992.

26. Michalczky, M.J.; West, R.; Michl, J. J. Am. Chem. Soc., 1984, 106, 821.

27. Watanabe, H.; Okawa, T.; Kato, M.; Nagai, Y. J. Chem. Soc., Chem. Commun., (1983), 781.

28. Masamune, S.; Tobita, H.; Murakami, S. J. Am. Chem. Soc., 1983, 105, 6524.

29. Watanabe, H.; Kougo, Y.; Nagai, Y. J. Chem. Soc., Chem. Commun., (1984), 66.

30. Ishikawa, M.; Kumada, M. Chem. Commun., (1970), 612 and (1971), 507.

31. Watanabe, H.; Kougo, Y.; Kato, M.; Kuwabara, H.; Okawa, T.; Nagai, Y.; Bull. Chem. Soc. Jpn, 1984, 57, in press.

CHAPTER 10

CHEMISTRY OF UNSTABLE INTERMEDIATES DERIVED FROM REACTIONS OF ORGANOHALOSILANES WITH ALKALI METAL VAPORS

L. E. Gusel'nikov, Yu. P. Polyakov, E. A. Volnina, and N. S. Nametkin
Topchiev Institute of Petrochemical Synthesis of the USSR Academy of Science,
Moscow, USSR

INTRODUCTION

Reactions of organohalosilanes with alkali metal vapors are of great interest in the gas phase study of organosilicon unstable intermediates such as silyl and silylmethyl radicals, silylenes, silenes, and disilenes. The vapor-phase method has an edge over the classic Wurtz method as the products can be quickly removed from the reaction zone. This is of particular importance when the desired products cannot be obtained because of secondary reactions initiated by alkali metals. Since the studies of Skell and co-workers [1-4], these reactions have not received much attention.

In this lecture we shall discuss the results of our recent studies on dehalogenation of a series of mono- and dichlorosilanes with K/Na alloy vapors; the studies were aimed at proving the participation of certain types of reactive intermediates and low stability small ring compounds containing silicon atoms [5-11]. Attention was focused on radical decomposition reactions, intramolecular rearrangements of silylenes, generation and reactions of silaalkenes, cyclization to prepare mono- and disilacyclobutane ring systems.

We shall base our discussion on the reactions of following compounds:

$$\underline{Me_2RSiCl} \ (\underline{1})$$

$\underline{1a}$: R=Me; $\underline{1b}$: R=Pr; $\underline{1c}$: R=Me$_3$Si; $\underline{1d}$: R=Me$_3$SiCH$_2$

$$\underline{ClCH_2(Me_2)SiR} \quad (\underline{2})$$

$\underline{2a}$: R = Me; $\quad \underline{2b}$: R = Me$_3$Si; $\quad \underline{2c}$: R = ClCH$_2$SiMe$_2$

$$\underline{MeRSiCl_2} \quad (\underline{3})$$

$\underline{3a}$: R = Me; $\quad \underline{3b}$: R = Pr; $\quad \underline{3c}$: R = Me$_3$SiCH$_2$;

$\underline{3d}$: R = Me$_3$SiCH$_2$CH$_2$; $\quad \underline{3e}$: R = Me$_3$SiCH$_2$CH$_2$CH$_2$

$$\underline{Cl(CH_2)_xSiMe(R)Cl} \quad (\underline{4})$$

$\underline{4a}$: x = 1, R = H; $\quad \underline{4b}$: x = 1, R = Me;

$\underline{4c}$: x = 3, R = H; $\quad \underline{4d}$: x = 3, R = Me

$$\underline{ClMe_2Si(CH_2)_xSiMe_2Cl} \quad (\underline{5})$$

$\underline{5a}$: x = 0; $\quad \underline{5b}$: x = 1; $\quad \underline{5c}$: x = 2; $\quad \underline{5d}$: x = 3

$$\underline{ClCH_2CH_2OSiMe_2Cl} \quad (\underline{6})$$

The reactions were carried out in a flow system.
Organochlorosilane vapors were introduced into the
reactor with a K/Na alloy(300-320°C, pressure below
1 Torr) and products were collected in a trap at 77K.

SILYL AND SILYLMETHYL RADICALS

Dehalogenation of $\underline{1}$ yields combination products
of corresponding silyl radicals:

$$Me_2RSiCl \xrightarrow{K/Na} Me_2RSi\cdot \longrightarrow Me_2RSiSiRMe_2$$

On dehalogenating the mixture of $\underline{1a}$ and $\underline{2a}$, we
obtained the main product

$$Me_3Si\cdot + \cdot CH_2SiMe_3 \longrightarrow Me_3SiCH_2SiMe_3$$

from combinations   of trimethylsilylmethyl and
trimethylsilyl radicals. Abstraction of H   by
silvl radicals is a less important reaction. Under
our experimental conditions, the decomposition
of silyl radicals was much less than in pyrolysis.
    The contribution of decomposition reactions of
silylmethyl radicals depends on the type of bonds at
the β position to the radical center. It is high for a
pentamethyldisilanylmethyl radical (βSi-Si bond)
and negligible for trimethylsilylmethyl radicals (β
Si-C bond). The latter produces a combination
product and tetramethylsilane. Below are shown the
reactions that occur upon dehalogenation of $\underline{2b}$ with
K/Na vapors:

$$Me_3SiSiMe_2CH_2Cl \xrightarrow{K/Na} Me_3SiSiMe_2\overset{\bullet}{C}H_2$$

$$\text{Me}_3\text{SiSiMe}_2\overset{\bullet}{\text{CH}}_2 \longrightarrow \text{Me}_3\text{Si}^{\bullet} + \text{Me}_2\text{Si}{=}\text{CH}_2$$

$$2\ \text{Me}_3\text{Si}^{\bullet} \longrightarrow \text{Me}_3\text{SiSiMe}_3\ ,\quad 17\%$$

$$2\ \text{Me}_2\text{Si}{=}\text{CH}_2 \longrightarrow \begin{array}{c}\text{Me}_2\text{Si}\text{---}\\ \text{---}\text{SiMe}_2\end{array}\ ,\quad 26\%$$

$$\text{Me}_3\text{Si}^{\bullet} + \text{Me}_2\text{Si}{=}\text{CH}_2 \longrightarrow \text{Me}_3\text{SiCH}_2\overset{\bullet}{\text{SiMe}}_2$$

$$\text{Me}_3\text{SiCH}_2\overset{\bullet}{\text{SiMe}}_2 \xrightarrow{\text{(H)}} \text{Me}_3\text{SiCH}_2\text{SiMe}_2\text{H}\ ,\quad 19\%$$

$$\begin{array}{c}\text{Me}_3\text{SiCH}_2\overset{\bullet}{\text{SiMe}}_2\\ \text{Me}_3\text{SiSiMe}_2\overset{\bullet}{\text{CH}}_2\end{array}\Big\rangle \xrightarrow{\text{Me}_3\text{Si}^{\bullet}} \text{Me}_3\text{SiCH}_2\text{SiMe}_2\text{SiMe}_3, 5\%$$

The alternative mechanism for the production of $\text{Me}_3\text{Si}^{\bullet}$ and $\text{Me}_2\text{Si}{=}\text{CH}_2$, involving isomerisation and decomposition of an ISO-radical:

$$\text{Me}_3\text{SiSiMe}_2\overset{\bullet}{\text{CH}}_2 \longrightarrow \text{Me}_3\text{SiCH}_2\overset{\bullet}{\text{SiMe}}_2$$

$$\text{Me}_3\text{SiCH}_2\overset{\bullet}{\text{SiMe}}_2 \longrightarrow \text{Me}_3\text{Si}^{\bullet} + \text{Me}_2\text{Si}{=}\text{CH}_2$$

is not significant as under our conditions 1d reacts yielding only the combination and H atom abstraction products.

SILYLENES

Dehalogenation of 3 with K/Na alloy vapors results in silylenes. By this method, we generated dimethylsilylene in excess buta-1,3-diene and obtained a good yield of 1,1-dimethyl-1-silacyclopent-3-ene:

$$\text{Me}_2\text{SiCl}_2 \xrightarrow{\text{K/Na}} \text{Me}_2\text{Si:} \longrightarrow \begin{array}{c}\text{Me}\\ \text{Me}\end{array}\!\!\diagdown\!\text{Si}\!\!\diagup\!\!\square$$

Without traps, dimethylsilylene forms polysilane oligomers [12] while alkylmethylsilylenes undergo also intramolecular C-H insertion. Among those, βC-H insertion is the most pronounced reaction, followed by a [3+2+1] -cyclodecomposition of the exited silirane intermediate:

$$\begin{array}{c}\text{R-CH-CH}_2\\ \text{H}\!\!\diagdown\!\!:\text{SiMe}\end{array} \longrightarrow \left[\begin{array}{c}\text{R-CH-CH}_2\\ \diagdown\!\!\diagup\\ \text{H}\diagup\text{Si}\diagdown\text{Me}\end{array}\right]^{*} \xrightarrow{\text{-MeHSi:}} \text{RCH}{=}\text{CH}_2$$

The intramolecular C-H insertion mechanism is confirmed by the isolation of both 1,3-disilacyclopentane (δ C-H insertion) and 1,3-disilacyclohexane (ε C-H insertion) in the dehalogenation of 3d and 3e, respectively:

The contribution of γ C-H insertion is negligible. Neither monosilacyclobutanes nor their decomposition products were obtained from alkylmethylsilylenes.

SILENES AND DISILENES

Dechlorination of 4a and 4b yields 1,3-disila-cyclobutanes via cyclodimerization of silaethylene intermediates:

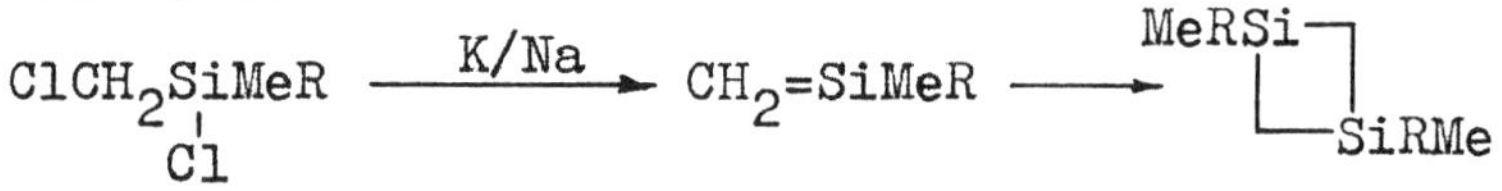

No products due to insertion of dimethylsilylene into Si-H bond were found. This indicates that 1-methyl-1-silaethylene does not rearrange into dimethylsilyl-ene. The absence of products with $SiCH_2CH_2Si$ group-ings indicates that $Me_3SiCH_2SiMe_2Cl$ is formed rather due to the addition of CH_2SiMe_2Cl radical to 1,1-di-methyl-1-silaethylene followed by H-atom abstraction, than by condensation involving C-Cl and Si-Cl bonds of 4 (x=1).

Dechlorination of 5a results in tetramethyldisi-lene and dimethylsilylene intermediates:

$$ClMe_2SiSiMe_2Cl \xrightarrow{K/Na} Me_2Si=SiMe_2$$

This is proved by isolation of 1,1,2,2-tetramethyl-1,2-disilacyclohex-3-ene and 1,1-dimethyl-1-sila-cyclopent-3-ene when the reaction is carried out in excess C_4H_6.

CYCLIZATION OF ORGANOHALOSILANES WITH K/Na VAPORS TO OBTAIN SMALL RING COMPOUNDS

Dehalogenation of 4c and 4d gives a good yield of monosilacyclobutanes while that of 6 does not yield 2-silaoxetane but its decomposition products: ethylene and trimer of dimethylsilanone.

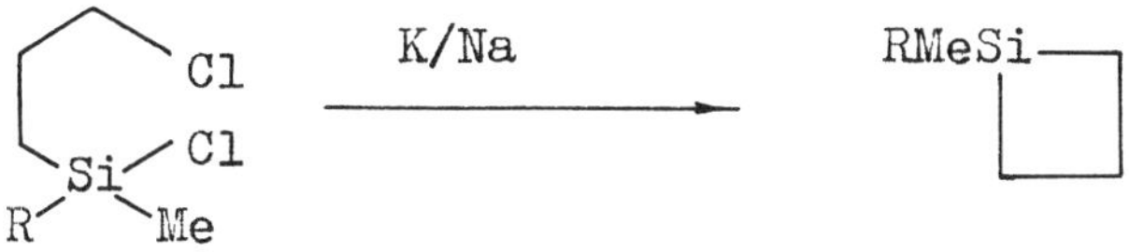

Cyclization of 5c give rise to 1,1,2,2-tetramethyl-1,2-disilacyclobutane - a low stability compound. It polymerizes at room temperature, but can be stored at 77K.

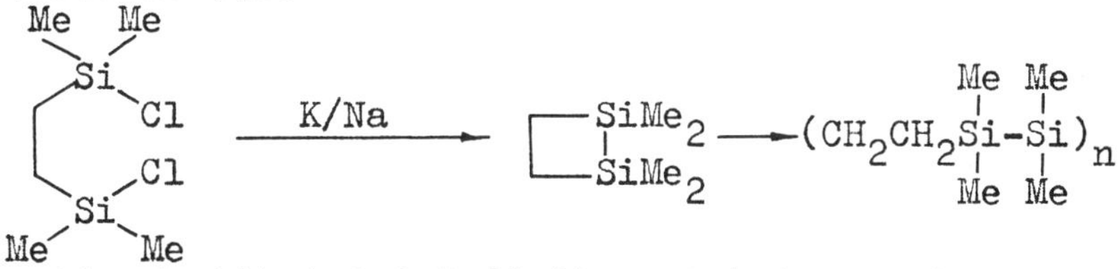

Other Si-substituted 1,2-disilacyclobutanes show similar properties.

We failed to obtain 1,2-disilacyclobutane from 2c presumably because of the decomposition of initial radicals resulting in 1,1-dimethyl-1-silaethylene and other reactive species. Thus, 1,1,3,3-tetramethyl-1,3-disilacyclobutane was isolated instead of the expected product:

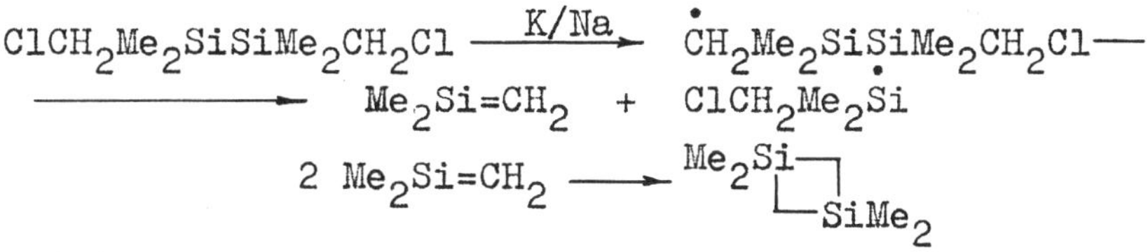

No 1,1,2,2-tetramethyl-1,2-disilacyclopropane but its dimer and the rearrangement product were obtained in dehalogenation of 5b. On the contrary, 5d gave a good yield of tetramethyl-1,2-disilacyclopentane.

CONCLUSION

From the discussion it is seen that various problems related to the chemistry of silyl radicals, silylenes, silaalkenes, as well as small ring silacycles may be solved by studing the dehalogenation of organochlorosilanes with K/Na vapors. Besides, the available starting materials, simple apparatus and the facilities for safe experimental work enable the organosilicon chemists to make a wider use of this method.

REFERENCES

1. Skell P.S.; Goldstein E.J. Silacyclopropanes.
   J. Am. Chem. Soc. 1964, 86, 1442.
2. Skell P.S.; Goldstein E.J. Dimethylsilylene.
   Ibid, 1442-43.
3. Skell P.S.; Goldstein E.J.; Petersen R.J.;
   Tingey G.L. Mono and polyradicals from compounds
   containing halogen. Chem. Ber. 1967, 100, 1442-
   46.
4. Skell P.S.; McGlinchey M.J. Cryochemistry (Mos-
   covits M.; Ozin G.A., Eds.), John Wiley & Sons,
   1976, p. 137-65.
5. Gusel'nikov L.E.; Polyakov Yu.P.; Nametkin N.S.
   Synthesis of strained cyclocarbosilanes with
   Si-Si bond by reactions of α,ω-bis(dimethylchlo-
   rosilyl) alkanes with alkali metal vapors. Dokl.
   Akad. Nauk SSSR. 1980, 253, 1133-36.
6. Gusel'nikov L.E.; Lopatnikova E.; Polyakov Yu. P.;
   Nametkin N.S. Generation and reactions of methyl-
   (trimethylsilylalkyl)silylenes in a gas phase.
   Dokl. Akad. Nauk SSSR. 1980, 253, 1387-89.
7. Gusel'nikov L.E.; Kerzina Z.A.; Polyakov Yu. P.;
   Nametkin N.S. A new route to 1,1-dimethyl-1-sila-
   cyclopent-3-ene. Izv. Akad. Nauk SSSR. Ser. Khim.
   1982, 219.
8. Gusel'nikov L.E.; Kerzina Z.A.; Polyakov Yu. P.;
   Nametkin N.S. A new reaction for formation of
   unstable 2,2-dimethyl-2-silaoxetane. Zh. Obshch.
   Khim. 1982, 52, 457-58.
9. Gusel'nikov L.E.; Polyakov Yu. P.; Zaikin V.G.;
   Nametkin N.S. Generation of 1,1-dimethyl-1-sila-
   ethylene by dehalogenation of chloromethyldime-
   thylchlorosilane with alkali metal vapors. Dokl.
   Akad. Nauk SSSR. 1984, 274, 598-602.
10. Gusel'nikov L.E.; Polyakov Yu. P.; Volnina E.A.;
    Ivanov A.V.; Nametkin N.S. Generation of 1,1-di-
    methyl-1-silaethylene in the reaction of chloro-
    methylpentamethyldisilane with alkali metal
    vapors. Izv. Akad. Nauk SSSR. Ser. Khim.(in press).
11. Gusel'nikov L.E.; Polyakov Yu. P.; Volnina E.A.;
    Zaikin V.G; Nametkin N.S. Synthesis of 1-methyl-
    1-silacyclobutane in the gas phase dehalogenation
    reaction of ɣ-chloropropylmethylchlorosilane with
    alkali metal vapors. Dokl. Akad. Nauk SSSR (in
    press).
12. Skell P.S.; Nefedov O.M. Formation and reactions
    of Me_2Si: and Me_2Ge: in a gas phase. Dokl. Akad
    Nauk SSSR. 1981, 259, 377-79.

CHAPTER 11

HIGHLY STERICALLY HINDERED ORGANOSILICON COMPOUNDS

C. Eaborn School of Chemistry and Molecular Sciences, University of Sussex, Brighton BN1 9QJ, U.K.

This account is concerned with compounds in which the very bulky $(Me_3Si)_3C$ ligand (the " trisyl" ligand, denoted by Tsi), or a related ligand, is attached to a silicon or other metal or metalloid centre. The novel properties to which such ligands give rise are due primarily to their bulk, but very interesting features often arise from the ease of elimination of Me_3SiX (e.g. X = halogen) of the type $(Me_3Si)_3CML_nX \rightarrow (Me_3Si)_2C=ML_n$.

<u>Compounds with Trisyl or Similar Ligands Attached to Elements other than Silicon</u>. The compounds described in this review were derived from trisyl-lithium or related species. Trisyl-lithium (for convenience denoted by TsiLi), made from $(Me_3Si)_3CH$ and MeLi in THF [1], is itself of interest since X-ray diffraction has shown it to have the structure (Ia), and so to be the first example of an organolithium ate species [2a]. Organolithium compounds are normally dimeric, trimeric <u>etc</u>., with the organic groups bridging between Li atoms; presumably the bulk of the Tsi ligand prevents such bridging, and the normally less favoured ate structure is adopted.

$[Li(THF)_4][M(Tsi)_2]$

M=Li, (Ia); Cu (Ib); Ag (Ic)

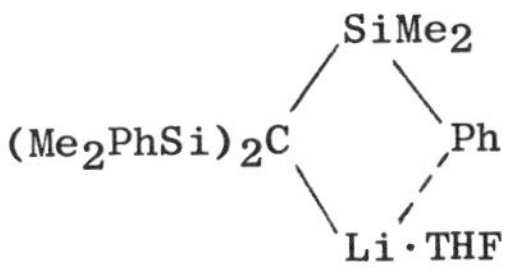

$(Me_2PhSi)_2C$ with SiMe$_2$ / Ph / Li·THF bridge

(TpsiLi)   (II)

Interestingly the lithium compound derived analogously from
(Me$_2$PhSi)$_3$CH (TpsiH) has the totally different and also novel
structure (II) [2b]. This reagent is significantly less reactive than TsiLi, (Ia); for example it does not react with
Me$_3$SiCl in refluxing THF, under conditions in which (Ia) reacts
quite rapidly [3]. It does couple, however, with the less
hindered Me$_2$SiHCl to give TpsiSiMe$_2$H.

Reactions of (Ia) with CuI and AgI give the corresponding
cuprate and argentate derivatives (Ib) [2d] and (Ic) [2c],
respectively, again the first examples of such species.

Treatment of TsiLi, (Ia), with the halides MCl$_2$, where
M = Zn, Cd, and Hg, give the compounds MTsi$_2$ [4]. These are
remarkably thermal stable (decomposing only above 300°C) and
chemically inert; for example, the zinc compound, although a
dialkylzinc species, can be steam distilled [4c].

Reactions of TsiLi with PR$_n$Cl$_{3-n}$ give TsiPR$_n$Cl$_{2-n}$ compounds;
such compounds with R = 0 or 1 have given rise to much new
phosphorus chemistry, including isolation of compounds of the
type (Me$_3$Si)$_2$P=CR (formed by eliminations of Me$_3$SiCl) and of
species containing P=P double bonds, e.g. TsiP=PTsi [5]. The
compound TsiPPh$_2$ is remarkable in that in MeOH it loses its
Me$_3$Si groups in stepwise fashion, finally to give MePPh$_2$; this
process, which is thought to involve initial protonation at
phosphorus is completely inhibited by the presence of a little
sodium methoxide [6a].

When TsiLi made in the usual way is treated with BF$_3$ or
AlCl$_3$, complications arise from the presence of THF, the main
products being TsiB[O(CH$_2$)$_4$Tsi]F and TsiAl[O(CH$_2$)$_4$Tsi]Cl; though
a little TsiAlCl$_2$ is also obtained [6b]. There is no such
complication with TpsiLi, (II), which with BF$_3$, for example,
gives TpsiBF$_2$; hydrolysis of the latter gives TpsiB(OH)F, the
first example of a compound with both a hydroxy and a halogeno
group on boron [6c].

The trisyl group has been attached to a range of other
elements; some leading references can be found in refs. [5(b)]
and [7].

Trisyl-Silicon and Related Compounds. It is possible here to
survey only a fraction of the chemistry which has been developed
for compounds in which a Tsi, Tpsi, or a related ligand is
attached to silicon bearing one or more functional groups. The
emphasis will be on the mechanism of substitution at such silicon centres, but it is appropriate first to mention some other
novel processes and species which have been observed:

(a) Thermolysis of TsiSiPh$_2$F, involving loss of Me$_3$SiF, gives

products which can be accounted for only by assuming rapid
isomerization of the initially formed $(Me_3Si)_2C=SiPh_2$ by
migrations of Me and Ph groups to give a mixture of this sila-
olefin with $(Me_3Si)(Ph_2MeSi)C=SiMe_2$, $(Me_3Si)(PhMe_2Si)C=SiPhMe$
and $(PhMe_2Si)_2C=SiMe_2$ [8a].

(b) Rearrangements involving novel 1,3-migrations of silyl
groups from C to O, e.g. $(Me_3Si)_2C(SiMe_3)(SiR_2OH) \rightarrow (Me_3Si)_2CH-$
(SiR_2OSiMe_3), can occur on treatment of suitable silanols with
base [8b]. For R = Ph the rearrangement occurs readily in
neutral MeOH, and the corresponding reaction in the case of
$C(SiMe_2OH)_4$ occurs readily on heating alone [8c].

(c) Compounds of the type $TsiSiMe_2OCOR$ react fairly readily
with NaOMe-MeOH, to give $TsiSiMe_2OH$ and MeOCOR, by cleavage of
the acyl-O bond, i.e. by the mechanisms well known for solvolysis of the corresponding alkyl esters but normally unobserv-
able for silicon esters [8d]. (The same process occurs with
$t-Bu_3SiOCOR$.)

(d) Treatment of $TsiSiR_2I$ compounds with AgOCN has given the
first silicon cyanates, e.g. $TsiSiMe_2OCN$ [8e,f]. The latter
undergoes isomerization to the isocyanate on heating (this
process in CCl_4 is greatly catalysed by ICl), and very readily
in MeOH if a little NaOMe is present. In MeOH alone solvolysis
is very fast, and cyanate appears to be an even better leaving
group than triflate (OSO_2CF_3).

The first normal thiocyanate of silicon, $(Me_3Si)_2C-$
$(SiMe_2OMe)(SiMe_2SCN)$, has recently been obtained by treatment
of $(Me_3Si)_2C(SiMe_2OMe)(SiMe_2Cl)$ with AgSCN [8g].

Substitution at Silicon. (a) Direct bimolecular substitutions.
The observation of unusual mechanisms of reaction at silicon
centres in trisyl-silicon compounds is made possible mainly by
the marked steric inhibition of direct bimolecular attack.
However, with suitable nucleophiles such substitutions do take
place, e.g. with F^-, N_3^-, CN^-, OCN^-, and SCN^-, in MeOH or
CH_3CN [9a]. Towards $TsiSiMe_2I$ the order of effectiveness of
various salts is $NaN_3, > CsF > KCN > KSCN > KOCN$ in MeOH, and
$KSCN > KOCN > KCN$, CsF in CH_3CN. (The reactions with KOCN and
KSCN give silicon isocyanates and isothiocyanates, respectively.)

(b) Reactions via silicocations. Iodides of the type $TsiSiR_2I$
are thought to generate Me-bridged cations of the type (III) on
treatment with silver or mercury(II) salts [9b,c]. When R = Ph
the products, e.g. from AgY in CH_2Cl_2, are exclusively the
rearranged species (IV) (R = Ph), which would be formed by
attack of Y^- at the less hindered Si(3) atom of (III) [9c].
When R = Et, for which there is less difference in hindrance at
Si(1) and Si(3), roughly equal amounts of unrearranged

(TsiSiEt$_2$Y) and rearranged products (IV, R = Et) are formed. Similar rearrangements take place when the reactions are carried out in MeOH, but the main products in this case are the silicon methoxides formed by attack of the solvent on the bridged cations (III).

Analogous migrations of the OMe group have recently been observed; e.g. treatment of (Me$_3$Si)$_2$C(SiMe$_2$OMe)(SiPh$_2$Br) with AgBF$_4$ in Et$_2$O gives predominantly the rearranged fluoride (Me$_3$Si)$_2$C(SiMe$_2$F)(SiPh$_2$OMe) [9d]. Calculations on the model bridged cations (V) indicate that the stabilization of the bridged relative to the open ion (H$_3$Si)$_2$C(SiMe$_2$Z)(SiH$_2^+$) increases in the sequence Me < Cl < F < OH < NH$_2$ [9e].

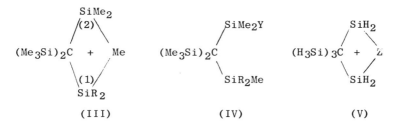

(III)   (IV)   (V)

Examples of other reactions involving Me migrations and thought to proceed through bridged cations of type (III) are (a) the solvolysis of TsiSiR$_2$I compounds in CF$_3$CO$_2$H [9c] (see later), and (b) the following reactions of TsiSiPh$_2$I, to give (Me$_3$Si)$_2$C(SiPh$_2$Me)(SiMe$_2$Y): (i) with ICl (Y = Cl) [9f]; (ii) with MeOH under UV irradiation (Y = OMe) [9g], and (iii) with MeOH in the presence of m-chloroperoxybenzoic acid (Y = OH) [9h].

1,3-Migrations of an Me group from Ge to Si have recently been observed; e.g. (Me$_3$Si)$_2$C(GeMe$_3$)(SiMe$_2$Br) reacts with AgBF$_4$ to give a mixture of (Me$_3$Si)$_2$C(GeMe$_3$)(SiMe$_2$F) and (Me$_3$Si)$_2$C(GeMe$_2$F)(SiMe$_3$) [9i].

(c) Solvolysis. The methanolyses of TsiSiMe$_2$X (X = OClO$_3$, OSO$_2$CF$_3$, I) and of TsiSiRHI (R = Ph or Me) were found not to be significantly accelerated by NaOMe, and so by the reasoning used for the classical assignment of an S$_N$1 mechanism to solvolysis of t-butyl halides, these methanolyses were assumed to be of the S$_N$1 type, involving rate-determining ionization to give a Me-bridged cation [10a]. This appeared to be confirmed by the observation that in the case of TsiSiPhHX species with X = Br, ONO$_3$, Cl, NCO, F etc., the reactions are accelerated by NaOMe [10a,b], thus establishing a dependence of mechanism in this system on the leaving group X. However, it was subsequently found that hydrolysis of TsiSiR$_2$I compounds (R = Ph or Et) and methanolysis of the compound with R = Et, gave only unrearranged products TsiSiR$_2$OH and TsiSiEt$_2$OMe, respectively, implying that symmetrical bridged cations of the type (III) are never present [10a]. The reactions may involve processes, possibly of the S_N2 (intermediate) type now thought to operate in solvolysis of t-butyl halides [10d], in which there is nucleophilic assistance by the solvent in the rate-determining step (leading to a nucleophilically solvated cation) though with considerable development of cationic character in the transition state. This would be consistent with the fact that TsiSiMe$_2$X compounds with X = OClO$_3$, OSO$_2$CF$_3$ or I do not react with the strongly electrophilic but weakly nucleophilic alcohol CF$_3$CH$_2$OH. With a sufficiently electrophilic solvent, CF$_3$CO$_2$H, an S_N1 process does appear to operate; thus TsiSiEt$_2$I gives roughly equal amounts of rearranged and unrearranged trifluoroacetates [9c].

Before leaving solvolysis of simple trisyl-silicon iodides it should be mentioned that the reaction of TsiSiPh(OH)I with MeOH on addition of a little NaOMe takes place much too readily to involve direct displacement, and the following novel silanone mechanism has been postulated [10e]:

$$
\underset{I}{\overset{OH}{Tsi(Ph)Si}} \quad \overset{HO^-}{\rightleftharpoons} \quad Tsi(Ph(Si \quad \rightarrow \quad Tsi(Ph)Si{=}O \quad \overset{MeOH}{\rightarrow} \quad \underset{OMe}{\overset{OH}{TsiPhSi}}
$$

There is clear evidence that certain groups, Z, can supply powerful anchimeric assistance to the leaving of a group X in reactions of (Me$_3$Si)$_2$C(SiMe$_2$Z)(SiMe$_2$X) species. Thus the compound with Z = OMe, X = Cl reacts with MeOH at least 10^6 times as rapidly as TsiSiMe$_2$Cl [11a], and it also reacts very rapidly with silver salts, towards which TsiSiMe$_2$Cl is inert [11a, 8g]. The obvious interpretation appeared to be that the MeO-bridged cation is formed very readily in the rate-determining step in both types of reaction, and the fact that the reaction with CF$_3$CH$_2$OH is in this case faster than that with MeOH is in keeping with this explanation. Surprisingly, however, the ethanolysis of (Me$_3$Si)$_2$C(SiMe$_2$OMe)(SiPh$_2$Br) appears to give only unrearranged product [9d], and it seems likely that in solvolysis even of (Me$_3$Si)$_2$C(SiMe$_2$OMe)(SiMe$_2$Cl), (VI), the bridged cation may never be fully formed. More puzzlingly, the presence of the γ-OMe group in (VI) markedly activates the Si-Cl bond towards bimolecular substitution by N$_3$, SCN$^-$ etc., though the effect is very much smaller than in the solvolyses [8g], and it seems that in the case of (VI) even this type of reaction involves development of substantial silicocationic character in the transition state. The most striking example of the effect of a γ-OMe group is provided, however, by the ready cleavage of an Si-Me bond, with generation of MeH, on treatment of TsiSiMe$_2$OMe with CF$_3$CO$_2$H at room temperature, TsiSiMe$_3$ being completely inert even under reflux [11a].

A γ-vinyl (Vi) group is also able to provide substantial anchimeric assistance, though much less effectively than a γ-OMe group. Thus $(Me_3Si)_2C(SiMe_2Vi)(SiMe_2I)$ (VII) reacts much more readily than $TsiSiMe_2I$ (VIII) with silver salts and with CF_3CO_2H, but in reactions with alkali metal salts or with MeOH the two iodides have rather similar reactivities [11a]. In reactions with alkali metal salts, (VII) is only slightly more reactive than (VIII), rather more in line with expectation.

Some interesting results have emerged from comparisons [12] of the reactions of $t-Bu_3SiI$ (IX) and $TsiSiMe_2I$ (VIII). (a) Under conditions in which (VIII) is thought to react via a fully-formed silicocation, namely solvolysis in CF_3CO_2H or reactions with silver salts, (IX) is much less reactive. (b) The reactivity difference between the two iodides is much smaller in reactions with alkali metal salts, (VIII) commonly being 3-8 times the more reactive. (c) In methanolysis and hydrolysis, in which a type of S_N2 (intermediate) process may operate, the reactivity differences vary considerably with the medium, with (VIII) usually, but not always, the more reactive, but again the differences are much smaller than those in the reactions mentioned under (a) in which the ease of ionization of the Si-X bond is the greatly dominant factor. Interestingly, in reactions with NaN_3, KOCN, and KSCN in CH_3CN, the chloride $t-Bu_3SiCl$ is more reactive than the iodide $t-Bu_3SiI$.

Much of the work from my group described above was jointly supervised by Dr.J.D. Smith. Support from the S.E.R.C. is gratefully acknowledged.

REFERENCES

1.   M.A. Cook, C. Eaborn, A.E. Jukes and D.R.M. Walton, J. Organomet. Chem., 1970, 24, 529; Z.K. Aiube and C. Eaborn, ibid, 1984, in the press.

2.(a) C. Eaborn, P.B. Hitchcock, J.D. Smith and A.C. Sullivan, J. Chem. Soc., Chem. Commun., 1983, 827; (b) idem, ibid, 1390; (c) idem, ibid, 1984, in the press; (d) idem, J. Organomet. Chem., 1984, 263, C23.

3.   C. Eaborn and A.I. Mansour, J. Organomet. Chem., in the press.

4.   (a) A.B. Bassindale, A.J. Bowles, M.A. Cook, C. Eaborn, A. Hudson, R.A. Jackson and A.E. Jukes, J. Chem. Soc., Chem. Commun., 1970, 559, (b) F.Glockling, N.S. Hosmane,

V.B. Mahale, J.J. Swindall, L. Magos and L.J. King, J. Chem. Res., (M) 1977, 1198; (c) C. Eaborn, N. Retta and J.D. Smith, J. Organomet. Chem., 1980, 190, 101.

5. For leading references see: (a) H. Schmidt, C. Wirkner and K. Issleib, Z. Chem., 1983, 23, 67; (b) A.H. Cowley, J.E. Kilduff, E.A.V. Ebsworth, D.W.H. Rankin, H.E. Robertson and R. Seip, J. Chem. Soc., Dalton Trans., 1984; 689, (c) J. Escudie, C. Couvret, H. Ranaivonjatovo and J. Satgé, Phosphorus and Sulfur, 1983, 17, 221.

6. (a) C. Eaborn, N. Retta and J.D. Smith, J. Chem. Soc., Dalton Trans., 1983, 905; (b) C. Eaborn, N.M. El-Kheli, N. Retta and J.D. Smith, J. Organomet. Chem., 1983, 249, 23; (c) N.M. El-Kheli, unpublished results.

7. N.N. Zemlyanski, I.V. Borisova, V.K. Bel'skii, N.D. Kolosova and I.P. Beletskaya, Bull. Acad, Sci., U.S.S.R., 1983, 869; F. Glockling and W.K. Ng, J. Chem. Soc., Dalton Trans., 1981, 1101.

8. (a) C. Eaborn, D.A.R. Happer, P.B. Hitchcock, S.P. Hopper, K.D. Safa, S.S. Washburne and D.R.M. Walton, J. Organomet. Chem., 1980, 186, 309; (b) R. Damrauer, C. Eaborn, D.A.R. Happer and A.I. Mansour, J. Chem. Soc., Chem. Commun., 1983, 348; (c) C. Eaborn, P.B. Hitchcock and P.D. Lickiss, J. Organomet. Chem., 1984, 264, 119; (d) R. Damja, C. Eaborn and A.K. Saxena, unpublished results; (e) C. Eaborn, P.D. Lickiss, G. Marquina-Chidsey and E.Y. Thorli, J. Chem. Soc., Chem. Commun., 1982, 1326; (f) C. Eaborn, Y.Y. El-Kaddar and P.D. Lickiss, J. Chem. Soc., Chem. Commun., 1983, 1450; (g) N.M. Romanelli, unpublished results.

9. (a) S.A.I. Al-Shali and C. Eaborn, J. Organomet. Chem., 1983, 246, C34; (b) C. Eaborn, J. Organomet. Chem., 1982, 239, 93; (c) C. Eaborn, D.A.R. Happer, S.P. Hopper and K.D. Safa, 1980, 188, 179; (d) S.T. Najim, unpublished results; (e) A.J. Kos, personal communication; (f) C. Eaborn and A.I. Mansour, J. Organomet. Chem., 1983, 254, 273 and refs. therein; (g) C. Eaborn, K.D. Safa, A. Ritter and W. Binder, J. Chem. Soc., Perkin Trans. 2, 1982, 1397; (h) A.L. Al-Wassil, C. Eaborn and A.K. Saxena, J. Chem. Soc., Chem. Commun., 1983, 974; (i) A.K. Saxena. unpublished results.

10. (a) C. Eaborn and F.M.S. Mahmoud, J. Chem. Soc., Perkin Trans. 2, 1981, 1309; (b) Z.I. Aiube, unpublished results; (c) S.A.I. Al-Shali, C. Eaborn, F.A. Fattah and S.T. Najim, J. Chem. Soc., Chem. Commun., 1984, 318; (d) T.W. Bentley and G.E. Carter, J. Am. Chem. Soc., 1982, 104, 5741; (e) Z.I. Aiube, J. Chojnowski, C. Eaborn and W.A. Stańczyk, J. Chem. Soc., Chem. Commun., 1983, 493.

11.   (a) C. Eaborn and D.E. Reed, J. Chem. Soc., Chem. Commun.,
       1983, 495; (b) C. Eaborn, P.D. Lickiss and N.A. Ramadan,
       J. Chem. Soc., Perkin Trans. 2, 1984, 267; (c) A.G. Ayoko,
       unpublished results.

12.   C. Eaborn and A.K. Saxena, J. Organomet. Chem., 1984, in
       the press.

CHAPTER 12

NEW RESULTS CONCERNING SILICON-TRANSITION METAL CHEMISTRY

R. J. P. Corriu Laboratoire de Chimie Organometalliques, Universite des Sciences et Techniques du Languedoc, Montpellier 34060, France

We describe a reinvestigation of transition metal complexes of π-silacyclopentadienyl (1-4). The preparation of these complexes was undertaken in the following three ways:

USE OF 2,5-DIPHENYLSILACYCLOPENTADIENE

Because of the possible preparation of functionnal siloles (2,5-diphenyl substituted) and symthesis of sila-anionic species, we have reexamined the synthesis of complexes by this way.

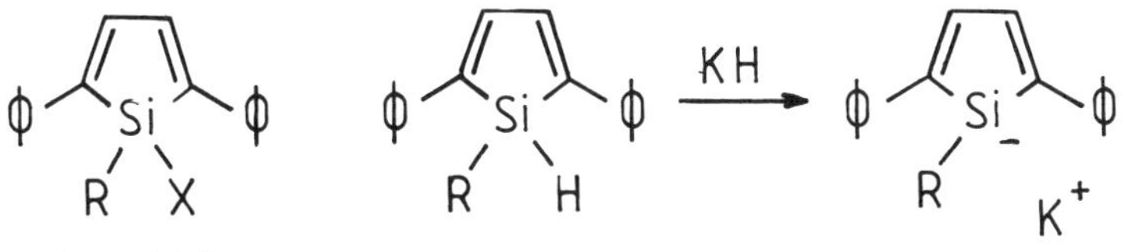

$(X)(X)$

$(R = Me, \phi ; X = H Cl)$ $(R = Me, vinyl)$

$(X = H, Cl)$

It is possible to obtain some complexes of the π-dienyl system and transition metal atoms with a functionnal silicon.

Some bimetallic complexes have also been obtained.

USE OF UNSUBSTITUTED SILOLES

1,1-dimethylsilole, reported by two groups (5,6), is a good precursor and some complexes have been obtained. The great stability of the π-diene system complexed with transition metals is very surprising (7).

The phosphine exchanges a CO group. The π-diene system is not displaced (7).

The cobalt carbonyl complex behaves simimarly to the π-cyclopentadienyl iron moiety (7).

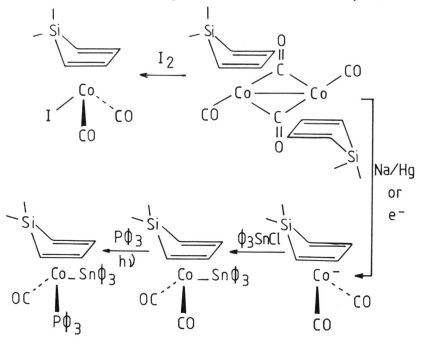

A bis 1,3-diene complex was also obtained (7).

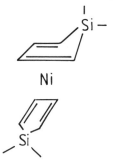

USE OF 3,4-DIMETHYLATED SILOLES (8)

Because of the very fast Diels-Alder reaction of the silole, previously used,   complexation must be performed at low temperatures (-70 to 0°C). The 3,4-dimethylated silole can be complexed at higher temperatures. It gives stable complexes.

$(M = Cr, Mo, W)$

The X-ray structure of the molybdenum complex has been studied. It shows an octahedral geometry. The variable ^1H NMR of Cr, Mo and W complexes shows a fluxionnality of these complexes attributed to motion involving the CO groups.

REFERENCES

1 - J.C. BRUNET, J. BERTRAND and C. LESENNE
    J. Organomet. Chem., 71 (1974) C8 and references therein.

2 - E.W. ABEL, T. BLACKMORE and R.J. WHITLEY
    J. Chem. Soc. Dalton Trans. (1976), 2484.

3 - H. SAKURAI, J. HAYASHI and T. KOBAYASHI
    J. Organomet. Chem., 110 (1976), 303.

4 - W. FINK
    Helv. Chim. Acta, 57 (1974), 167.

5 - A. LAPORTERIE, P. MAZEROLLES, J. DUBAC and H. ILOUGHAME
    J. Organomet. Chem., 206 (1981), C25 and 216 (1981) 321.

6 - G.T. BURNS and T.J. BARTON
    J. Organomet. Chem., 209 (1981) C25.

7 - G.T. BURNS, E. COLOMER and R.J.P. CORRIU
    Organometallics, 2 (1983) 1901.

8 - Cooperation with J. DUBAC, A. LAPORTERIE and H. ILOUGHAME
    (Toulouse) Cf A. LAPORTERIE, G. MANUEL, J. DUBAC and P.
    MAZEROLLES, Nouv. J. Chim., 6 (1982) 67.

CHAPTER 13

ORGANOSILICON AND RELATED ORGANO-NONMETALLIC SPECIES: LIGAND EFFECTS ON STEREOCHEMISTRY AND REACTIVITY

J. C. Martin, William H. Stevenson III, and David Y. Lee Department of Chemistry, Roger Adams Laboratory, University of Illinois, Urbana, IL 61801, USA

INTRODUCTION

The systematic development [1] of qualitative structure-reactivity relationships for derivatives of hypervalent [2] nonmetals has allowed us to design ligands which are very effective at stabilizing hitherto unknown species of novel structural types. As might be expected, the incorporation of structural features known to provide strong stabilization of a given hypervalent species (e.g., the bidentate ligands of phosphorane **2** [3] into isoelectronic, isostructural species such as siliconate anion **1** [4], or the cationic persulfonium ion **3** [5], provides extraordinary stabilization to these species as well.

1 (10-Si-5)   **2** (10-P-5)   **3** (10-S-5)

Compounds **1**, **2**, and **3** are labelled using the N–X–L designation, in which N valence shell electrons are designated as being formally involved in bonding L ligands to the central atom X [6].

The two bidentate ligands attached to the silicon of 8-Si-4 silane **4** are known [1] to be very effective in stabilizing 10-X-5 species such as **1**, **2**, or **3**. The three-center four-electron hypervalent [2] bond of trigonal bipyramidal (TBP) species, such as phosphorane **2**, is highly polarized, with relatively negative charge on the apical ligands and relatively positive charge on the central atom. The effective electronegativity of the apical oxygens of **2** is enhanced by the inductive effect of the trifluoromethyl groups and the equatorial, relatively electropositive aryl sigma-donor substituents stabilize the relatively positive charge on the phosphorus atom of **2**.

STABILIZED 10-Si-5 SPECIES

4 (8-Si-4)   **6** (10-Si-5)

The effect of this bidentate ligand on the reactivity of silane **4** is predictable in that it makes **4** a strong Lewis acid which reacts with very weak nucleophiles such as, e.g., benzaldehyde **5** to give 10-Si-5 species **6**. The stabilizing influence of this ligand set on the TBP siliconate is sufficient to make **6** directly observable in solution at low temperature by NMR.

While the ^{19}F NMR of silane **4** in a hydrocarbon solvent shows the two multiplets expected for the nonequivalent geminal CF_3 groups, the addition of a small amount of a nucleophilic catalyst such as benzaldehyde results in coalescence of the CF_3 peaks into a sharp singlet at room temperature. The rate of the reaction producing the coalescence of the CF_3 peaks, an inversion of configuration at 8-Si-4 silicon, is found to be first order in added nucleophile [4]. (The rates were determined either by NMR line shape analysis or by magnetization transfer methods.) This is in sharp contrast to the observations of Corriu [7] that racemizations of a number of optically active silanes are catalyzed by added nucleophiles in a reaction second (or higher) order in nucleophile. This is explained by the reversible formation of the achiral octahedral 12-Si-6 intermediate or transition state, **7**.

$$\text{2 Nu}^{\bar{}} \qquad\qquad \mathbf{7} \ (12\text{-Si-6}) \qquad\qquad \text{2 Nu}^{\bar{}}$$

The first order kinetics for the racemization of silane
is explained by invoking rapid inversion of configuration at
10-Si-5 silicon by nondissociative permutational isomerization.
Five Berry pseudorotation steps interconverting TBP geometries
can be chained together to provide a route for inversion of
configuration at silicon [4,8,9].

The five-step pseudorotation process for inversion of configuration at TBP silicon has been established as a likely
route for the observed inversion by finding a dependence of the
rate of inversion on the apicophilicity of the nucleophile Y,
pictured in **8** as occupying an equatorial position. This was
investigated for a variety of equatorial ligands, Y, including
several in which it could be shown that the Si-Y bond was not
broken (Y = n-butyl [4], aryl [4], or fluorine [9] during the
inversion. The rate was found to be well correlated with
$\sigma^+ (\rho^+_{422K} = 0.33)$.

This is consistent with the postulated five-step pseudorotation mechanism for inversion via a transition state resembling the high energy TBP geometry **9**. Ring strain in the
diequatorial five-membered ring of **9**, and the apical location
of its electropositive carbon ligand, combine to make it a high
energy species which is probably very near to the transition
state for inversion. The energetic cost of attaining a transition state geometry similar to **9** can be paid in part by the
shift of ligand Y from an equatorial position, in ground state
8, to an apical position in **9**. If Y is very apicophilic, as it
is in **6**, its shift to the apical position in transition state **9**
is energetically favored.

9

The value of ΔG^*_{424K} for TBP inversion ranges from 28.6 kcal/mol for **8**, Y = n-butyl, [8] to 17.5 kcal/mol for Y = F [9]. The barrier for nondissociative inversion of **6** (**9**, with $-Y = -O^+\!\!=\!\!=\!\!CHAr$) is calculated from estimated substituent effects [8] to be in the range of 10 kcal/mol. This reaction is therefore rapid enough to account for the observed rate of inversion of silane **4** catalyzed by benzaldehyde **5**.

The bidentate ligands of **8** stabilize this 10-Si-5 species to allow catalysis of the inversion of 8-Si-4 silane **4** by very weak nucleophiles in a process first-order in nucleophile. No evidence is seen for a second-order process for inversion (via **7**), consistent with the idea that the bidentate ligand does not provide specific stabilization for the octahedral 12-Si-6 species. Since the six ligand sites of an octahedral species are identical, it is not expected that the specific stabilization provided to a TBP species by a bidentate ligand with one σ-donor and one σ-acceptor site could be as effective in an octahedral species.

AN UNCATALYZED NONDISSOCIATIVE MECHANISM FOR INVERSION OF 8-Si-4 SILANES

Since weak nucleophiles catalyze the inversion of **4** via TBP 10-Si-5 siliconates it is not surprising that the weakly nucleophilic sites (e.g., the oxygens) of silane **4** can make possible its self-catalyzed inversion. When the rate of inversion of **4** is followed by ^{19}F NMR at 422K in decahydronaphthalene-d_{18} over a concentration range from 0.018 to 0.22 M the pseudo first-order rate constant (k_{obs}) varies linearly with the concentration from 0.48 ± 0.01 to 2.45 ± 0.5 s^{-1} expected for a second-order reaction ($k_2 = 9.8 ± 2.6$ $Lmol^{-1}s^{-1}$). Extrapolation to infinite dilution gives evidence for a residual first-order rate constant $k_1 = 0.30 ± 0\text{-}11$ s^{-1} ($\Delta G^*_{422K} = 26.1 ± 0.3$ kcal/mol).

$$k_{obs} = k_1 + k_2 \quad [4]$$

The most probable mechanism for the first-order process inverting silane **4** is the nondissociative inversion via the planar 8-Si-4 silicon species. The energy difference between distorted tetrahedral geometry of the silane **4** and its planar geometry **10** is therefore equal to the observed ΔG^*, 26.1 kcal/mol.

$$4 \rightleftarrows \left[\begin{array}{c} F_3C \\ F_3C \end{array} \underset{O}{\overset{O}{\diagdown}} Si \overset{O}{\underset{}{\diagup}} \begin{array}{c} CF_3 \\ CF_3 \end{array} \right] \rightleftarrows \overline{4}$$

10

Theoretical calculations of square planar silicon species have suggested [11] that their energies can be not too far above the energy of the tetrahedral geometry. Indeed, at one point X-ray crystallographic evidence was interpreted [12] in terms of a coplanar ground state geometry for the silane with two catecholate bidentate ligands attached to silicon, a tetraaryloxy 8-Si-4 species. This interpretation was, however, later refuted [13].

Structural features predicted [11] to favor the planar geometry for 8-Si-4 species, namely σ-acceptor, π-donor ligands, are present in the bidentate ligands of **4**. All four ligand sites are occupied by π-donor ligands (aryl or alkoxy) and two are occupied by good σ-acceptors, the fluoroalkoxy ligands.

The X-ray structure of **4** [8] shows an O-Si-O angle of 113.4° and endocyclic C-Si-O angles of 94.2° and 94.6°. These distortions of angles from the tetrahedral values are in the direction of the planar geometry of **10**, and suggest that the relief of angle strain in the five-membered ring, on going to the 90° C-Si-O angles of the square planar geometry of **10**, provides driving force for the pictured inversion mechanism.

OCTAHEDRAL 12-B-6 SPECIES

Silicon is more electropositive than carbon since it is below it in the periodic table. Boron, to the left of carbon, is also more electropositive. If the role of d-orbitals in stabilizing 10-X-5 and 12-X-6 species is, as much evidence indicates [1], less important than it has traditionally been considered to be, and if the stability of the hypervalent three-center bond is greater when there is a larger difference in electronegativity between the central atom and its apical ligands, then both hypervalent silicon and hypervalent boron species should be more stable than their carbon analogues.

We have reported [14] spectroscopic evidence for a 10-C-5 species, a hypervalent [2] carbon ground state analogous to the S_N2 transition state. More recently we have obtained evidence [15], from ^{11}B NMR and electronic spectra, for several 10-B-5 species which are more stable, being isolable as stable crystalline solids.

In view of the accessibility of 12-Si-6 species, and the above argument for possible similarities between hypervalent boron and hypervalent silicon, we thought it worthwhile to attempt to make a stable 12-B-6 analogue. The pictured reaction gave **11**, an isolable species for which a variety of spectroscopic evidence supports the postulated octahedral, 12-B-6 structure in solution.

R = t-Bu   **11** (12-B-6)

Evidence for the structure of **11** includes ^{19}F NMR (a singlet at temperatures as low as -150 °C), ^{13}C and ^{1}H NMR (peaks consistent with the symmetry and charge distribution in **11**) and mass spectrometry. The most compelling evidence, however, came from the ^{11}B NMR chemical shift for **11** (-122.9 ppm from BF -etherate). All known 8-B-4 borates with first-row elements attached to boron show ^{11}B chemical shifts more than 100 ppm downfield of this 12-B-6 species (in the range from +50.7 to -17.5 ppm [16]. The reported [15] 10-B-5 species (**12**, **13**, and **14**) also show ^{11}B NMR chemical shifts upfield of the 8-B-4 species, but downfield of the 12-B-6 species **11**, in the range -20.1 (for **12**), -35 (for **13**), and -41.0 ppm (for **14**). Protonation of **11** with one equivalent of trifluoromethanesulfonic acid converts it to a species with a chemical shift of -70.1 ppm, in a range near that of the 10-B-5 species.

12   R = t-Bu   **13**   **14**

In addition to evidence parallel to that for 12-B-6 species **11**, structural data for the 10-B-5 species includes another type of evidence, from the electronic spectrum of **12**. It is a yellow solid, also yellow in solution (λ_{max} = 397 nm, ε = 1650). The expected delocalization of an electron pair from the three-center four-electron O-B-O bond into the equatorial π-system makes **12** a bis-ipso-aromatic [14] analogue of the fluorenyl anion. Monoprotonation of **12** apparently occurs at oxygen to give an 8-B-4 species, probably **15**, showing two ^{19}F NMR singlets, which is colorless, consistent with more localized bonding at 8-B-4 boron. The stabilization expected from the π-acceptor equatorial ligands of **12** and **13** is not required to make 10-B-5 species observable, however. Compound **14**, which lacks this feature, shows a ^{11}B chemical shift at even higher field than **12** and **13**.

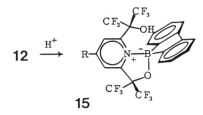

CONCLUSION

These observations suggest a substantial similarity between hypervalent silicon and boron species, consistent with the notion that the role of d-orbitals in the stabilization of hypervalent species has traditionally been overstated. The design of ligands which provide ligand-centered orbitals to accommodate the formal octet expansion of 10-electron and 12-electron species [1] can make accessible not only extremely stable hypervalent derivatives of second-row and third-row nonmetals, but also unprecedented hypervalent first-row organo-nonmetallic species.

REFERENCES

1. J. C. Martin and E. F. Perozzi, Science 1976, 191, 154;
    J. C. Martin, Ibid. 1983, 221, 509; J. C. Martin and W.
    H. Stevenson III, Phosphorus and Sulfur 1983, 18, 81.

2. J. I. Musher, Angew. Chem., Int. Ed. Engl. 1969, 8, 54.

3. A. B. Fiduccia, B. L. Murphy, and J. C. Martin, to be
    published.

4.   E. F. Perozzi, R. S. Michalak, G. D. Figuly, W. H. Steven-
     son III, D. B. Dess, M. R. Ross, and J. C. Martin, J. Org.
     Chem. 1981, 46, 1049; W. H. Stevenson III and J. C. Mar-
     tin, J. Am. Chem. Soc. 1982, 104, 309.

5.   R. S. Michalak and J. C. Martin, J. Am. Chem. Soc. 1980,
     102, 5921; R. S. Michalak and J. S. Martin, Ibid. 1982,
     104, 1683.

6.   C. W. Perkins, J. C. Martin, A. J. Arduengo, W. Lau, A.
     Alegria, and J. K. Kochi, J. Am. Chem. Soc. 1980, 102,
     7753.

7.   R. J. P. Corriu, G. Dabosi, M. Martineau, and C. J.
     Guerin, J. Organomet. Chem. 1980, 186, 25.

8.   W. H. Stevenson III, S. Wilson, and J. C. Martin, J. Am.
     Chem. Soc. Submitted for publication; W. H. Stevenson
     III and J. C. Martin, Ibid. Submitted for publication.

9.   W. B. Farnham and R. L. Harlow, J. Am. Chem. Soc. 1981,
     103, 4608.

10. W. J. Stevenson III and J. C. Martin, J. Am. Chem. Soc.
     Submitted for publication.

11. E.-U. Würthwein and P. v. R. Schleyer, Angew. Chem., Int.
     Ed. Engl. 1979, 18, 553; J. Chaudrasekhar, E.-U. Würth-
     wein, and P. v. R. Schleyer, Tetrahedron 1981, 37, 921.

12. H. Meyer and G. Nagorsen, Angew. Chem., Int. Ed. Engl.
     1979, 18, 551.

13. J. D. Dunitz, Angew. Chem., Int. Ed. Engl. 1980, 19, 1034.

14. T. R. Forbus, Jr. and J. C. Martin, J. Am. Chem. Soc.
     1979, 101, 5057.

15. D. Y. Lee and J. C. Martin, J. Am. Chem. Soc. In press.

16. H. Nöth and B. Wrackmeyer, "Nuclear Magnetic Resonance
     Spectroscopy of Boron Compounds", Springer-Verlag, New
     York, 1978.

CHAPTER 14

NEW ASPECTS OF SILACYCLOHEPTATRIENE CHEMISTRY

Yasuhiro Nakadaira Department of Chemistry, Faculty of Science, Tohoku University, Sendai 980, Japan

INTRODUCTION

The seven-membered cyclic triene with a hetero atom has received much attention due to its possible aromatic or antiaromatic character depending on the nature of the hetero atom contained in the ring[1]. Since the silicon is tetra-valent and has both donor and acceptor properties, it is interesting to know how its unique electronic features affect the nature of the silicon analogue of cycloheptatriene, silacycloheptatriene(silepin). So far, synthesis of silacycloheptatrienes has been limited to benzo-fused and C-substituted systems[2]. We would like to describe the first synthesis and some chemistry of one of the most fundamental silepin, 7,7-dimethyl-7-silacyclohepta-1,3,5-triene and some related compounds. The chemistry of the new type of silacycloheptatriene, isobenzosilepin is also described.

SYNTHESIS OF SILACYCLOHEPTATRIENE

Up to the present, several kinds of benzosilepins have been prepared by bromination with NBS followed by dehydrobromination as illustrated in eq(1) [2a]. Recently, the first synthesis of non-annulated silepin was accomplished by photoaddition followed by several steps as shown in eq(2)[2d].

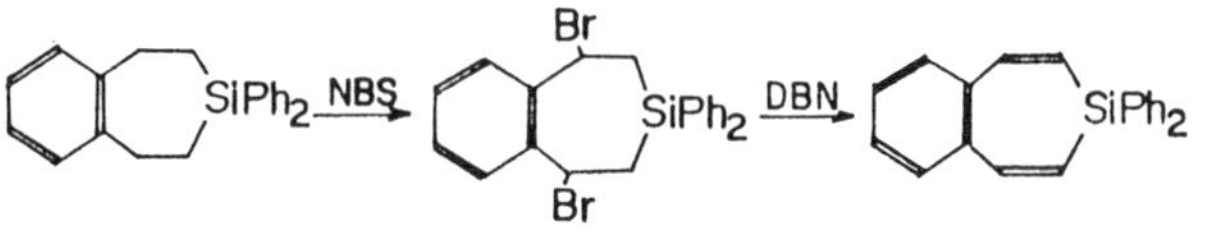

(1)

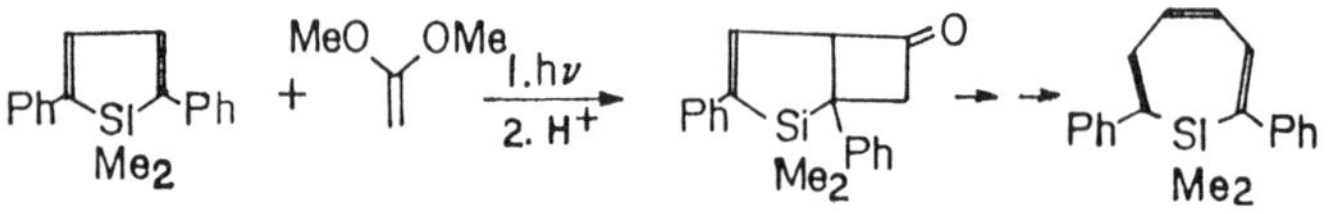

(2)

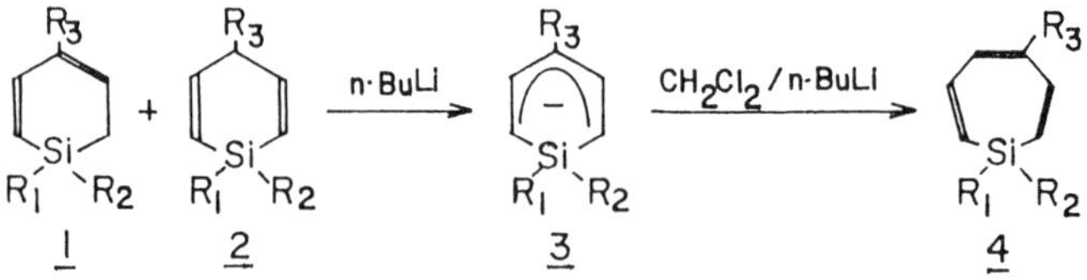

a, $R_1 = R_2 = Me$, $R_3 = H$; b, $R_1 = Me$, $R_2 = Ph$, $R_3 = H$,

c, $R_1 = $ 2-naphthyl, $R_2 = Me$, $R_3 = H$; d, $R_1 = R_2 = Me$, $R_3 = SiMe_3$.

Our new synthetic approach is based on allowing silacyclohexadienyllithium to react with dichloromethane in the presence of n-butyllithium. In a typical experiment, a mixture of silacyclohexadienes 1a and 2a(1a/2a = 60/40)[3] in dry ether was treated with excess amounts of n-butyllithium(3.4 equiv) at 0 °C in an argon atmosphere, and then a solution of dichloromethane(2.0 equiv) in dry ether was added dropwise at -78°C. The reaction mixture was allowed to warm to room temperature gradually, quenched with water and extracted with ether. After bulb to bulb distilation, the distillate obtained was purified by preparative GLC to give 7,7-dimethyl-7-silacycloheptatriene(4a) as a colorless oil in 20% yield[4]. Silepin 4a is well characterized by various spectroscopic evidences. In particular, the ^{1}H-NMR spectrum showed an [AA'BB'CC'] signal due to the six ring protons at δ 5.78 ~ 6.89 as shown in Figure I. Similarly, some related non-annulated silepins, 4b ~ d were prepared from corresponding silacyclohexadiene 1 and 2.

The treatment of silacyclohexadienyllithium 3a with dichloromethane in the presence of tetramethylethylenediamine at -78°C in ether afforded 4,4-dimethyl-4-silacyclohexa-2,5-dien-1-ylidene(5))[5]. This suggests that the direct coupling reaction be-

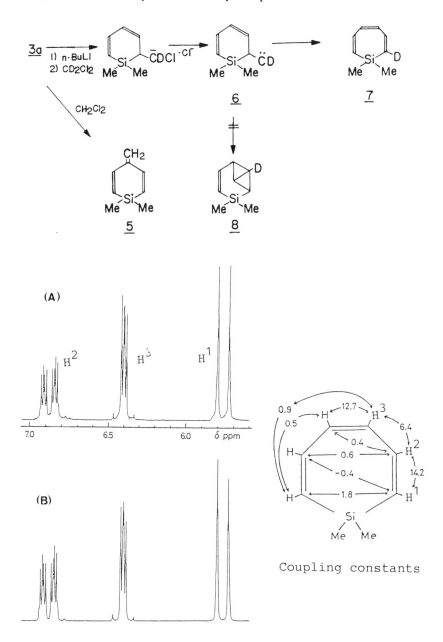

Figure I — Experimentals (A) and simulated (B) ¹H-NMR spectra of **4a** in CDCl₃, and coupling constants of the ring protons (Hz).

Table I – Coupling constant and dihedral angle (α) of cycloheptatriene and some heteropins.

compound	coupling constant $J_{2,3}$(Hz)	dihedral angle (degree)
cycloheptatriene	5.51[6a]	29.5[a,6b]
tropone	8.36[c]	0[b,6d]
2,7-di-t-Bu-thiepin	5.5[1b]	28.0[b,6e]
thiepin 1,1-dioxide	6.96[6e]	22.8[b,6f]
1-H azepine	5.43[6g]	28.1[b,c,6h]
7,7-di-Me-silepin **4a** (this study)	6.4	~25 (estimated)

a: Determined by microwave. b: Determined by X-ray crystallography. c: The value of 1-(p-bromophenyl-sulfonyl)-1H-azepine.

tween **3a** and dichloromethane is not involved in the reaction process to silepin **4a**. Then, **3a** was treated with dideuteriodichloromethane in the presence of n-butyllithium as described above, and resultant silepin 7 is found to be labelled exclusively at C(1). This finding indicates that bicyclobutane inter-mediate **8** is not involved in the formation of silepin **7**. On the base of these results, it can be safely concluded that the mechanism for the formation of silepin **7** involves the initial formation of key intermediate carbene **6** by the reaction of 3a with chlorocarbene followed by elimination of chloride ion, and then a preferential migration of the silyl group to the carbene center of **6** gives silepin **7**. This accords with the fact that under these conditions only 7,7-dimethyl-1,4-diphenylsilepin was obtained from 6,6-dimethyl-1,4-diphenyl-6-silacyclo-hexa-1,3-diene.

The ^{1}H-NMR spectra of **4a** remain invariant and signals due to Si-Me groups appear as a singlet over the substantial range, 30 ~ -122°C. Therefore, NMR evidence gives us no indication of the presence of

silanorcaradiene **9a** in equilibrium with silacycloheptatriene **4a**. Since except tropone, heteropins have
been shown to take a boat conformation, this should
be accounted for not by a planar structure, but by
rapidly inverting boat geometries. Although, to
present, no X-ray structural information has been
obtained in these silicon-containing six-π-electron
systems, conformational changes may be influential in
vicinal couplings between protons attached to sp^2
centres connected by a single bond($J_{2,3}$). Since the
relationship of Karplus appears to hold for vicinal
coupling in medium rings[7], the magnitude of $J_{2,3}$
should increase as the dihedral angle α between the
base plane[C(1), C(2), C(5), C(6)] and the stern
plane[C(2), C(3), C(4), C(5)] decreases. The data
listed in Table I allow to estimate that the dihedral angle α of silepin **4a** is around 25°. Consequently, a boat conformation of **4a** is inferred from
the analogy with those of other non-planar seven-
membered trienes listed in Table I.

THERMAL REACTIONS OF SILACYCLOHEPTATRIENE

Upon mild heating of **4a**, dimethylsilylene is
eliminated with the formation of benzene, probably
through silanorcaradiene **9a**. This is substantiated by
the experimental fact that the thermolysis of **4a** in
toluene containing 10 molar excess of 2,3-dimethyl-
butadiene gave dimethylsilylene trapped product,
1,2,4,4-tetramethyl-4-silacyclopentene(**10**) in 24%
yield[8]. On the other hand, on the thermolysis of
4a without solvent at 110°C for 15 h the silylene
once formed with benzene undergoes formal 1,4-addi-
tion with **4a** to afford 4,4,8,8-tetramethyl-4,8-di-
silabicyclo[3.2.1]octa-2,6-diene(**11**) in 25% yield .
On pyrolysis, some silepin, such as 7,7-dimethyl-1,6-
diphenyl-7-silacycloheptatriene[2d] and 4,5-benzo-
7,7-diphenyl-7-silacycloheptatriene[2a] have been
reported to evolve the corresponding silylene but
they require more forced reaction conditions, namely,
the reaction temperature reported is 250°C for the
former and 500°C for the latter.
Kinetic information for the thermolysis was
obtained by heating toluene-d_6 solutions of **4a** containing tetramethoxysilane as a silylene trap, and

following the disappearance of ^{1}H-NMR signal due to SiMe group of **4a** as a function of time. The reaction was found to be unimolecular , and kinetic parameters were obtained($\Delta H^{\ddagger}$ = 25.9 kcal/mol, and $\Delta S^{\ddagger}$ = -7.3 e.u.). This type of extrusion reaction has been known to occur readily in some cycloheptatriene ana- logues, such as thiepin, phosphepin, and borepin [1]. Tropone(eq 3) and its ethylene ketal(eq 4) also have been reported to give benzene on pyrolysis with ki- netic parameters ($\Delta H^{\ddagger}$ = 51.2 kcal/mol, and $\Delta S^{\ddagger}$ = 1.2 e.u. for tropone[9a], and $\Delta H^{\ddagger}$ = 30.8 kcal/mol, and $\Delta S^{\ddagger}$ = -7.2 e.u. for its ketal[9b]). Since these extrusion reactions proceed probably via norcaradiene intermediates, which are thought to extrude the cor-

responding fragment to give benzene quite readily, these kinetic parameters are reasonably attributed to those of ring closure. Tropone is a planar molecule and is stabilized through delocalization of six-π- electrons. On the other hand, silepin and the tropone ketal are non-planar cyclictrienes and are well suited for intramolecular cyclization.

PHOTOCHEMISTRY OF SILACYCLOHEPTATRIENE

Photochemical behaviours of cycloheptatriene and its analogue highly depend on the nature of the ring atom and substituents, and is affected by muluti- plicity of the excited state[1, 10]. Irradiation of bezene-d$_6$ solution of **4a** under argon with medium pressure Hg arc lamp through Pyrex well gave only complex mixtures and none of the photoproducts could be identified. However, irradiation of **4d** under similar conditions led to the disappearance of the silepin and the concomitant formation of 4-silabi- cycylo[3.2.0]hept-2,6-diene **12a** and **12b** in the ratio of 65 : 35. The photo-isomerization of **4d** was not effected by the presence of piperylene. This result suggests that **12a** and **12b** are formed from singlet excited state of **4d**. The photochemical behaviours of silepin seems to be similar to those of cyclo- heptatriene and its analogue, such as azepine, oxepin and tropone[1a, 10].

DIELS-ALCER REACTIONS OF SILACYCLOHEPTATRIENE

The addition of a dienophile to cycloheptatriene and its analogues has been known to be highly de- pendent on the nature of the atom contained in the ring. Thus, cycloheptatriene reacts with a dienophile such as 4-phenyl-1,2,4-triazoline-3,5-dione(**13**) to give the tricyclic adduct **14** formally derived from norcaradiene[11], and similarly oxepin gives the tricyclic epoxide with maleic anhydride[12]. On the other hand, tropone undergoes direct Diels-Alder reaction to two of the three double bonds of the triene[13]. Silepin **4a** is reluctant to undergo Diels- Alder reaction, for example **4a** did not react with bistrifluoromethylacetylene at ambient temperature, but decomposes to benzene with loss of dimethylsilyl- ene under more drastic conditions. However, **4a** re- acted with **13** to give (4 + 2)adduct **15** in dichloro- methane at room temperuture in 63% yield. The struc-

ture of the adduct can be derived readily from its
^{1}H- and ^{13}C-NMR spectra. Under similar conditions 3-
trimethylsilylsilepin **4d** afforded two types of (4 +
2)adduct, **16a** and **16b** in 57% yield (**16a**/**16b** = 4/1).
Since silacyclopropane and cyclopropanone are much
more strained in comparison with cyclopropane and
oxirane, in Diels-Alder reaction it is quite reasonable to expect that silepin and tropone should give
only usual (4 + 2) products such as **15**, not tricyclic
adduct as the cases of cycloheptatriene and oxepin.

TRANSITION METAL COMPLEXES OF SILACYCLOHEPTATRIENE

The reaction of cycloheptatriene with hexacar-
bonylmolybdenum gives tricarbonyl(η^6-cycloheptatri-
ene)molybdenum[14], and they have been converted to
the corresponding tricarbonyltropylium cation complexes[15].

Treatment of **4a** with hexacarbonylmolybdenum(1.5
equiv) in heptane for 7h at 100°C under argon atmosphere gave tricarbonyl(η^6-7,7-dimethyl-7-silacyclo-
heptatriene)molybdenum(**17**) as red plates in 28 %
yield[16]. Similarly, reaction of **4b** with hexacarbo-
nylmolybdenum(1.5 equiv) gave a mixture of **18a** and
18b in 9.8%(**18a**/**18b** = 3/1). The stereochemistry of
substituents on silicon are assigned from chemical
shifts of the SiMe signal.

CHEMISTRY OF ISOBENZOSILACYCLOHEPTATRIENE

A silanorbornadiene derivative has been known to act as a thermal and photochemical silylene precursor [17]. Recently, we have prepared a variety of 2,3-benzo-1,4-diphenyl-7-silanorbornadienes by the reaction of the corresponding 2,5-diphenyl-1-silacyclopentadiene with benzyne. On thermolysis and irradiation, these compounds are shown to generate the corresponding silylene quite efficiently[18]. Current interests in kinetically stable disilene have been focused on a highly crowded silylene[19]. However, it is difficult to prepare the benzosilanorbornadiene suited for a crowded silylene precursor by the method above. Although a few 2,3,5,6-dibenzosilanorborna-

diene have been reported[19], but their chemical properties remain unknown. The dibenzosilanorbornadiene with bulky substituents on silicon atom is stable and was obtained by the reaction of lithium anthracenide with the corresponding dihalosilane in THF, for example, dimesityldichlorosilane and t-butyl(2,4,6-triisopropylphenyl)difluorosilane gave the corresponding dibenzosilanorbornadiene 19a and 19b in 58 and 45% yields, respectively[20]. Irradiation of the benzene solution of 19 in a vacuum-sealed NMR tube generated the corresponding silylene which readily dimerized to form kinetically stable disilene 20[22]. Along with the generation of the silylene, 19 has turned out to isomerize photochemically to 21 by way of isobenzosilepin 23. 2,3,6,7-Dibenzo-4-sila-bicyclo[3.2.0]heptadiene 21 is fully compatible with its spectroscopic properties. Especially in the case of **21a**, in addition to two kinds of signals due to mesityl groups on the silicon atom, it shows two doublets(δ 3.61(d, J = 1.5 Hz), and 4.45(d, J = 1.5 Hz) in CD_2Cl_2), which are assignable to those of H(5) and H(1), respectively. This photo-induced isomerization can be rationalized as shown in Scheme. At first, **19** undergoes photoinduced 1,3-silylmigration to yield vinyl-silacyclopropane **22**. This type of 1,3-silylmigration has been observed in the photolysis of 5,6-disilacyclohexa-1,3-diene[23]. A vinyl-silacyclopropane is generally formed by addition of a silylene to a diene, and in certain circumstances [5,24], it is known to undergo a carbon-carbon bond fission instead of a carbon-silicon bond fission and is transformed to isobenzosilacycloheptatriene **23**. This contains an o-xylylene unit and undergoes ring closure to **21** quite readily. Attempts to trap **23** have been unsuccessful at this stage.

ACKNOWLEDGEMENT   It is a great pleasure to acknowledge the invaluable advice, stimulation and encouragement of Professor Hideki Sakurai. I also wish to thank my coworkers, Ryuji Sato, and Kazuya Oharu. Only their patience and skill made this work possible.

REFERENCES AND NOTES

1)   (a) L.A.Paquette, "Nonbenzenoid Aromatics," Vol. 1, J.P.Synder, Ed., Academic Press, New York, N.Y. 1970, p 249. (b) K.Yamamoto, S.Yamazaki, Y.Kohashi, I.Murata, Y.Kai, N.Kanehisa, K.Maki, and N.Kasai, Tetrahedron Lett., **23**, 3195(1982). (c) S.M.Van der Kerk, J.Boersma, and G.J.M.Van der Kerk, J.Organomet. Chem., **215**, 303(1981). (d) G.Markl and W. Burger, Tetrahedron Lett., 24, 2545(1983).

2) (a) L.Birkofer, and H.Haddard, Chem.Ber., **102**,
    432(1969);**105**,2101(1972). (b) T.J.Barton, W.E.
    Volz, and J.L.Johnson, J.Org.Chem., **36**, 3365
    (1971); J.Y.Corey, M.Dueber, and B.Bichlmeir,
    J.Organomet.Chem., **26** ,167(1971); E.K.Cartledge,
    and P.D.Mollere, ibid., **26**, 175(1971).
    (d) T.J.Barton, R.C.Kippenhan,Jr., and
    A.J.Nelson, J.Am.Chem.Soc., **96**, 2272(1974).
    (e) M.Ishikawa, T.Fuchikami, and M.Kumada,
    Tetrahedron Lett., 1299(1976).

3) R-J.Hwang, R.T.Conlin, and P.P.Gaspar, J.Organo-
    met.Chem., **94**, C38(1975).

4) **4a**: a colorless oil; MS m/e(%) M^+ 136(35); ^{1}H-
    NMR($CDCl_3$) δ 0.09(s, 6H, SiMe), 5.78(m, 2H, H^1,
    H^6), 6.89(m, 2H, H^2, H^5), 6.41(m, 2H, H^3, H^4)
    (Determined by spectral simulation); ^{13}C-NMR
    ($CDCl_3$) δ -3.00(q), 131.21(d), 132.12(d),
    140.41(d); ^{29}Si-NMR($CDCl_3$) δ -17.2; UV(n-hexane,
    nm(ϵ)) λsh 216(5300), λmax 281(2100).

5) Quenching of **3a** with trimethylchlorosilane af-
    forded 6,6-dimethyl-3-trimethylsilyl-6-sila-
    cyclohexa-1,4-diene in 80% yield(see, E.A.
    Chernyshev, N.G.Komlenkova, S.A.Bashkirova, A.V.
    Kisin, F.M.Smirnova, and V.A.Kironov, Zh.Ohshch.
    Khim., **44**, 226(1976), Chem.Abst., **80**, 96087j
    (1974).

6) (a)H.Günter, M.Gortotz, and H.Meisenheimer, Org.
    Mag.Resonace, **6**, 388(1974). (b) S.S.Butcher,
    J.Chem.Phys., **42**, 1833(1967). (c) D.J.Bertelli,
    T.G.Andrews,Jr., and P.O.Grews, J.Am.Chem.Soc.,
    91, 5286(1969). (d) M.J.Barrow, O.S.Mills, and
    G.Filippini, J.Chem.Soc.Chem Commun., 66(1973).
    (e) M.P.Willamson, W.K.Mock, and S.Castellano,
    Org.Mag.Resonace, **2**, 511(1970). (f) H.L.Ammon,
    P.H.Watts,Jr., J.M.Stewart, and M.L.Mock, J.Am.
    Chem.Soc., **90**, 45011968). (g) E.Vogel,
    H-J.Altenbach, J-M.Drossard, H.Schmickler, and
    Stegelmeoer, Angew.Chem.Int.Ed.Engl., **19**, 1061
    (1980). (h) I.C.Paul, S.M.Johnson, L.A.
    Paquette, J.H.Barrett, and R.J.Haluka, J.Am.
    Chem.Soc., **90**, 5023(1968).

7) M.Karplus, J.Am.Chem.Soc., **85**, 2870(1963).

8) W.H.Atwell, and D.R.Weyenberg, J.Am.Chem.Soc.,
    90, 3438(1968).

9) (a) A.Amano, T.Mukai, T.Nakazawa, and K.

Okayama, Bull.Chem.Soc. Japan, **49**, 1671(1977).
(b) T.Fukunaga, T.Mukai, Y.Akazaki, and R.
Suzuki, Tetrahedron Lett., 2975(1970).

10) D.J.Pasto, "Organic Photochemistry," Ed., O.L.
Chapman, Marcel Dekker, New York, N.Y., 1967, p.
155.

11) K.Alder, and G.Jacobs, Ber., **86**, 1528(1953).

12) E.Vogel, W.A.Boll, and H.Gunther, Tetrahedron
Lett., 609(1965).

13) T.Nozoe, T.Mukai, T.Nagase, and Y.Toyooka,
Bull.Chem.Soc. Japan, **33**, 1247(1960).

14) E.W.Abel, M.A.Bennett, R.Burton, and
G.Wilkinson, J.Chem.Soc., 4559(1958).

15) a) H.Dauben,Jr., and L.R.Honnen, J.Am.Chem.
Soc., **80**, 5570(1958). (b) J.D.Munro, and P.L.
Pauson, J.Chem.Soc., 3475(1961).

16) **17:** mp 75~76 °C; MS m/e(%) M^+ 318(24); ^{1}H-
NMR($CDCl_3$) δ -0.49(s, 3H), 0.57(s, 3H), 3.14
(d, 2H, J = 11.3 Hz), 5.7 ~ 5.9(m, 4H); ^{29}Si-
NMR($CDCl_3$) δ -7.4.

17) P.P.Gasper, "Reactive Intermediates," Vol. 1,
M.Jones,Jr., and R.A,Moss, Ed., John Wiley &
Sons, New York, N.Y.,1976, p. 229.

18) (a) H.Sakurai, H.Sakaba, and Y.Nakadaira,
J.Am.Chem.Soc., **104**, 6156(1982).
(b) H.Sakurai, Y.Nakadaira, and H.Sakaba,
Organometallics, **2**, 1484(1983).

19) (a) R.West, Pure and Appl.Chem., **56**, 163(1984).
(b) S.Murakami, S.Collins, and S.Masamune,
Tetrahedron Lett., **25**, 2131(1984).

20) B.Mayer and W.P.Neumann, Tetrahedron Lett., **21**,
4887(1980); We thank for Professor W.P.Neumann
for sending their unpublished results.

21) **19a:** mp 195 ~ 196°C; MS m/e(%) M^+ 444(48); ^{1}H-
NMR($CDCl_3$) δ 1.98(s, 6H), 2.33(s, 12H), 4.08(s,
2H), 6.4 ~ 7.2(m, 12H); ^{29}Si-NMR(CDCl3) δ 45.4.

22) The formation of disilene **20** is supported by
appearance of characteristic ^{29}Si-NMR signal at
δ 63.6(from **19a**) and 96.5(from **19b**), respectively.

23) Y.Nakadaira, S.Kanouchi, and H.Sakurai, J.Am.
Chem.Soc., **96**, 5621(1974).

24) H.Sakurai, Y.Kobayashi, R.Sato, and Y.Nakadaira,
Chem.Lett., 1147(1983).

CHAPTER 15

NEW DI AND POLYSILYLATED MODELS

J. Dunogues, C. Biran and M. Laguerre Laboratoire de Chimie organique du Silicium et de l'Etain, associé au C.N.R.S. (U.A. n° 35), Université de Bordeaux I, F 33405 Talence, France

INTRODUCTION

Successive disilylations of aromatic or haloaromatic substrates using organometallic routes ($\equiv$SiCl/Li/THF or $\equiv$SiCl/Mg/HMPA reagent) constitute an easy approach to highly polysilylated models [1], e.g. [2] :

The reaction proceeds from the formation of radical-anions (or dianions) with subsequent silylation [1] :

In the case of haloaromatic derivatives, substitution of chlorine (by SiMe$_3$) accompanies the additive disilylation.

Here we report some applications of this reaction in the preparation of organosilicon synthons, original stereochemical models or novel skeletons.

RESULTS

1 - An extensive study of the silylation of 1,2-dichlorobenzene has been carried out [3] and this substrate was found a versatile precursor of 1,2-, 1,3- or 1,4-bis ; 1,2,4-tris ; 1,2,4,5-tetrakis(trimethylsilyl)-benzenes ; 3,6-bis(trimethylsilyl)1,4-cyclohexadiene ; 2,3,5,6-tetrakis(trimethylsilyl)1,3-cyclohexadiene [3] , 1,3,4,5,6 pentakis(trimethylsilyl)-and 1,2,3,4,5,6 hexakis(trimethylsilyl)cyclohexenes [4]. The last was quantitatively formed upon tetrasilylation of 1,2-bis(trimethylsilyl) benzene [4]. Results are summarized in Scheme 1 :

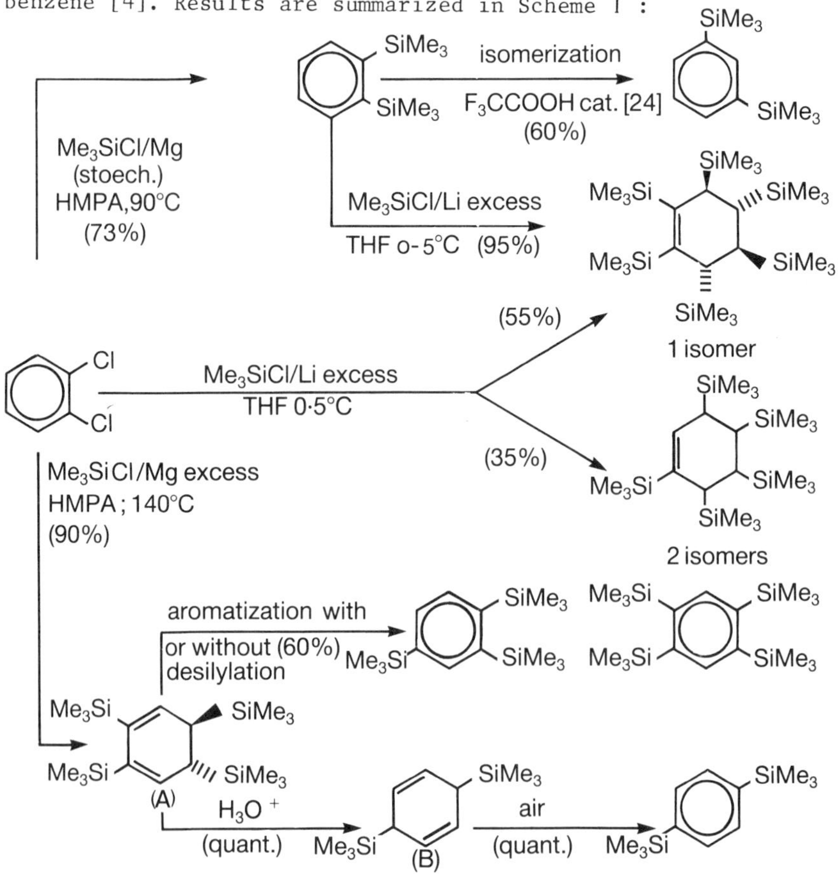

Scheme 1 — Versatile polysilylation of 1,2-dichlorobenzene.

These results require the following comments :

i. The hexasilyl derivative was constituted by only one isomer. Although we have not obtained at this time convenient
crystals for an X-ray structure analysis, NMR data allowed us to
assign to this compound the structure in which the four Me₃Si
groups bonded to sp³ hybridized carbon atoms are in *pseudo*-
axial position :

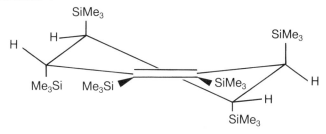

NMR data were found to be in good agreement with
that of 1-tBu,2,3,4,5-tetrakis(trimethylsilyl)1-cyclohexene resulting from *tetra*silylation of *tertio*butylbenzène (78 % yield)
[4]. In this case as well as for   pentasilyl cyclohexene
(2 isomers) the substitution of one Csp² by a tBu or Me₃Si
groups implies a *pseudo*-axial position for the Me₃Si group attached to the vicinal Csp³. The obtention of only one isomer in
the tetrasilylation of 1,2-bis(trimethylsilyl)benzene and *tertio*
butylbenzene and only two isomers in the tetrasilylation of
PhSiMe₃ (quant. yields [4]) is due to the high steric hindrance
in these derivatives.

In the case of the *tertio*butyl derivative (determination
by X-ray analysis) and for one isomer of the pentasilyl one
(determination by NMR comparison with the previous), the observed structure was the following :

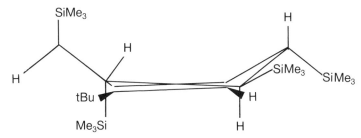

Concerning the second pentasilyl isomer a structure in good
agreement with the NMR data and the steric considerations was
proposed [4].

ii. Tetrasilylation of 1,2-dichlorobenzene under drastic
conditions afforded 2,3,5,6-tetrakis(trimethylsilyl)1,3-cyclo-
hexadiene (A), a new compound , which was realized to be an
attractive intermediate for synthesis:

- Upon hydrolysis it gave quantitatively 3,6-bis(trimethyl-silyl)1,4-cyclohexadiene (A). The obtention of this compound can be explained in terms of the formation of the more stabilized carbonium ion intermediate accompanied by the desilylation which involves a maximal steric decompression. As previously observed [5] desilylation from (B) is considerably slower, so (B) could be quantitatively isolated.

- (A) was found reactive in Diels-Alder reactions. Results are summarized in Scheme 2 :

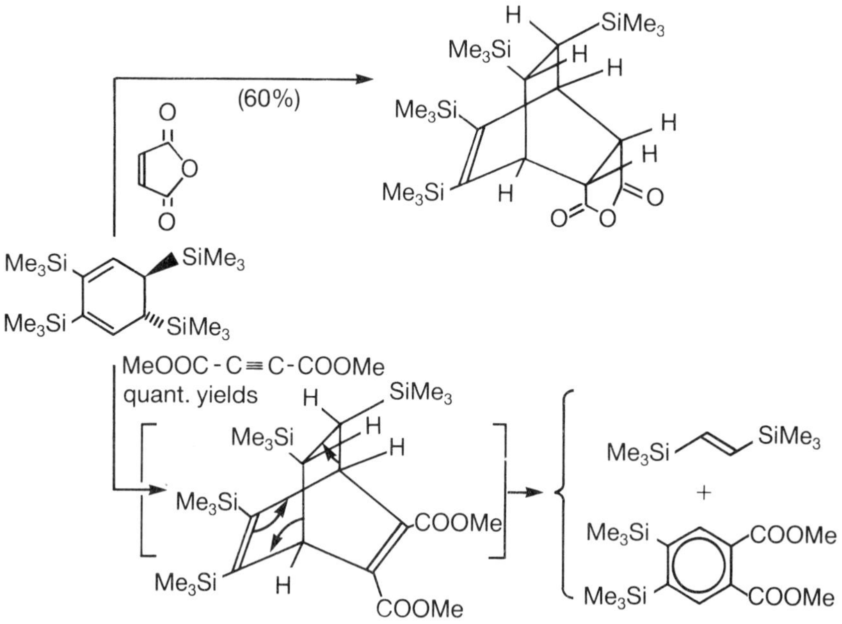

Scheme 2 — Behaviour of 2,3,5,6-tetrakis(trimethylsilyl)-1,3-cyclohexadiene in Diels-Alder reactions.

(C) was used in the synthesis of 4,5-diiodoorthophthalic acid [3].

iii) In previous papers we have reported the synthetic uses of (B) [6,7]. We also have to recall the synthetic potentialities of all the polysilylated aromatic derivatives reported in Scheme 1 since electrophilic substitution of the silyl group in these derivatives (ipso effect) opens wide perspectives in the polyfunctional benzene series [8,10].

2 - We have reinvestigated the persilylation of most of the polychlorobenzenes (so the two isomers of the pentasilylated cyclohexene could be obtained from chlorobenzene or 1,3- or 1,4-

dichlorobenzene [11]), but the most significant results were
obtained with hexachlorobenzene since we have optimized the
observations of GILMAN [12] and obtained quantitatively the
tetrakis(trimethylsilyl)allene   [13] :

$$C_6Cl_6 \xrightarrow[\substack{\text{large excess} \\ \text{0-5°C}}]{Me_3SiCl/Li/THF}$$

Me₃Si, SiMe₃ / Me₃Si, SiMe₃ (D) (95%)

Desilylation with 2 equiv. of $F_3C\text{-}COOH$ allowed to form
$Me_3Si\text{-}CH_2\text{-}C{\equiv}C\text{-}SiMe_3$ in quantitative yields. Although the study
of progressive desilylation from (D) is actually in progress,
1,3,3-tris(trimethylsilyl)1-propyne could be obtained in moderate yields upon monodesilylation of (D) [13] as well as tri-
methylsilylallene from 1,3-bis(trimethylsilyl)propyne.

3 - Silylation of pyrene with $Me_3SiCl/Mg/HMPA$ or
$Me_3SiCl/Li/THF$ was a very complex reaction. Among the formed
products we could identify after separation by HPLC [14] :

(E)   (F)   (G)

(H)   (I)

These preliminary results show the organosilicon route offers new possibilities for the functionalization of the skele-
ton (electrophilic substitution usually occurs in positions
1,3,6 and 8). Moreover (G) constitutes an original and attractive structure and (H) probably came from (I) (not isolated).

4 - Di- or polysilylation have been carried out in the acenaphthylene series [15]. Among the results we have to note the
special structure of the 1,2-bis(trimethylsilyl)acenaphthylene
obtained from additive disilylation of acenaphthylene followed
by oxidization (treatment by BuLi followed by action of $CdCl_2$)
[14] , in which the crystal structure shows a twisted double
bond :

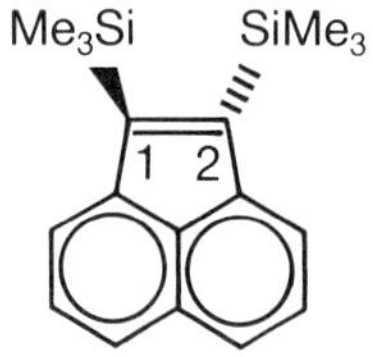

$$\widehat{Si_1C_1,C_2Si_2} \approx 17°$$

5 - Also we here report results obtained in the silylation
of heterobicyclic derivatives.

°Silylation of <u>benzofuranne</u> using an excess of Me_3SiCl and
Li in THF at 25°C afforded two main products [16] (at 0° only
a disilylation of the carbon-carbon double bond was observed) :

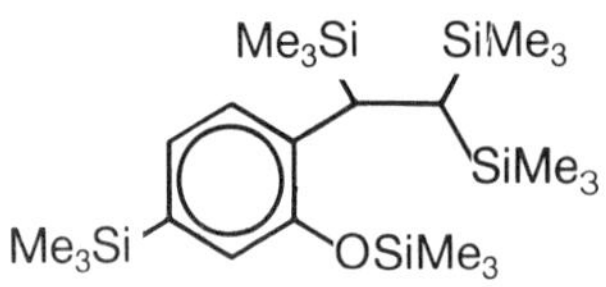

(J) (60%)

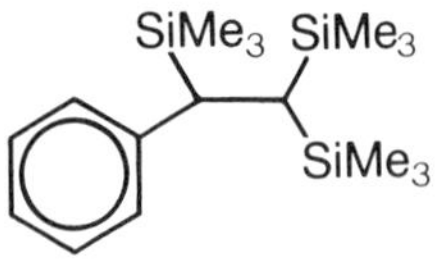

(K) 25%

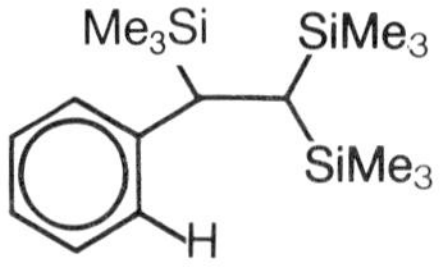

(L)

The structure of (J) which afforded (L) upon acidic hydrolysis in homogeneous medium, was confirmed by X ray analysis. In
the three cases, the NMR data confirmed the free rotation around
the C_7-C_8 bond is sterically hindered at room temperature. This
observation has to be compared with what we have observed with
tetrakis(trimethylsilyl)ethane [17] and bis(trimethylsilyl)ami-
no bis (or tris) trimethylsilyl methane [18,19] which exhibit
favored conformations at room temperature.

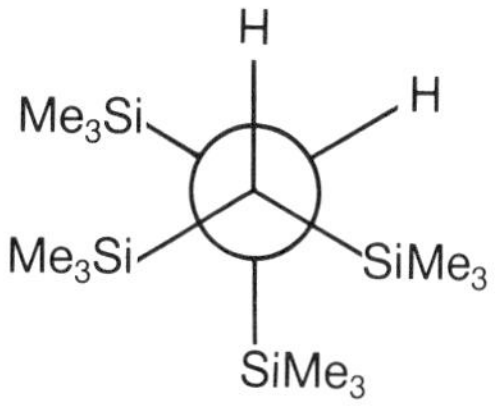

(M) 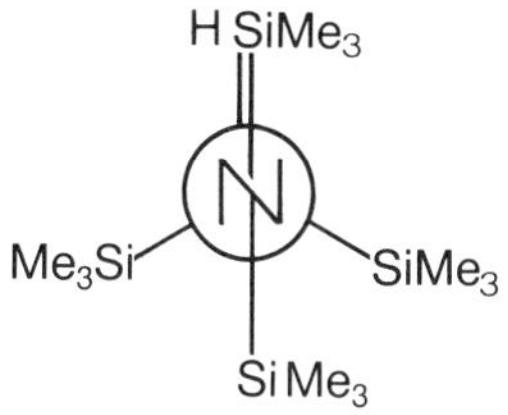   (N)

$(Me_3Si)_3C-N(SiMe_3)_2$

(study in progress)

(O)

In the case of (M) the *anti* is less stable than the *gauche*
form probably because of the repulsion of the trimethylsilyl
groups directly attached to each of the tertiary carbons.

The rotation around the carbon–carbon or carbon–nitrogen
bond can be observed by NMR spectroscopy when (M) or (N) is
warmed ((O)was not tested).

°Silylation of benzothiophene with the $Me_3SiCl/Li/THF$
reagent at 0-2°C provided (P) in 60 % yields [16] :

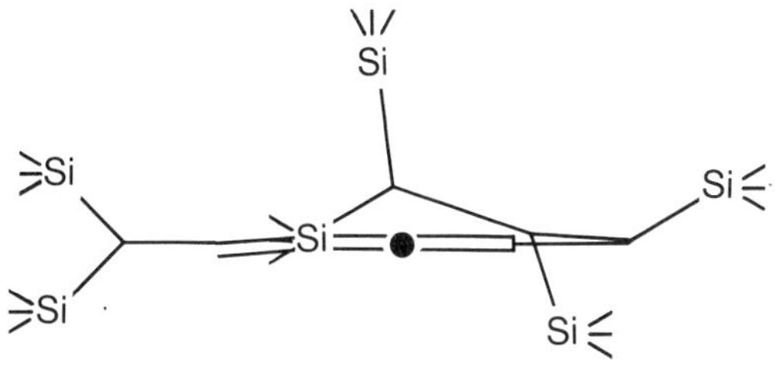

(P) 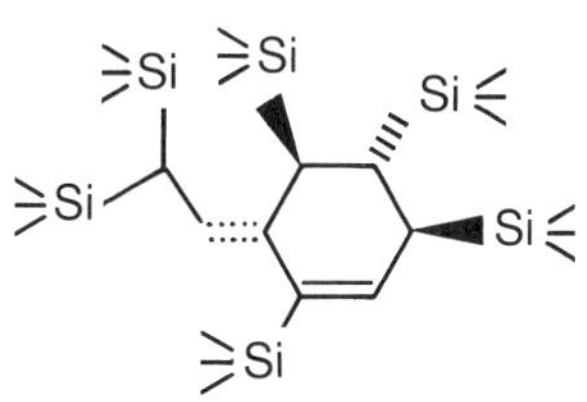   (P)

The structure of (P) was established by X-ray analysis.

In the case of dibenzothiophene among the formed products, $(Me_3Si)_2S$ (50 %) and

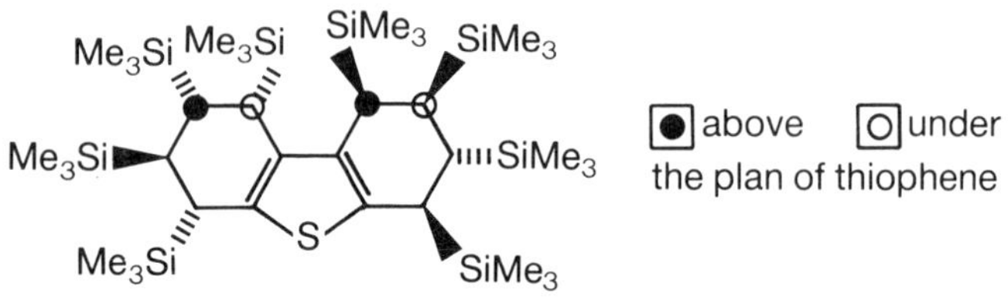

●above   ◯under
the plan of thiophene

(P)

were identified. The formation of $(Me_3Si)_2S$ shows the reductive silylation is accompanied by partial desulfurization [20].

°Silylation of indole afforded (Q) (80 % after cryst.) :

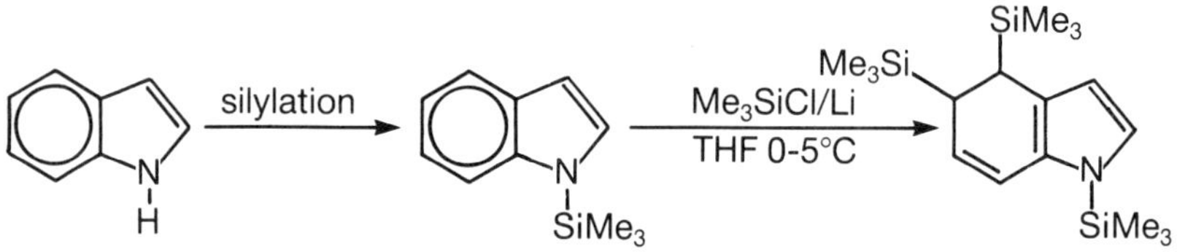

(Q) (80%)

The reaction gave a vicinal (and not 1,4- as previously reported by others [21]) disilylation on the aromatic ring. [Q] afforded (R) by aromatization using benzoquinone in 70 % yield, accompanied by (S) (30 % yield) :

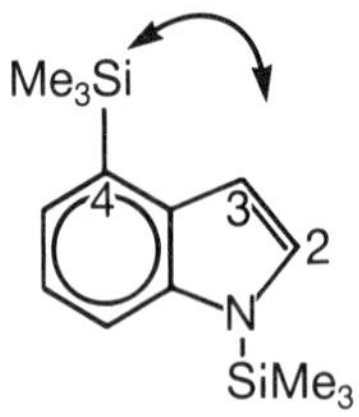

(R) (70%) 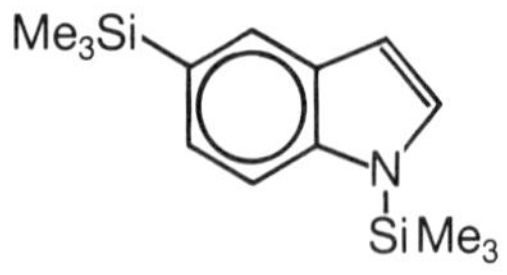   S(30%)

Considering the high reactivity of the position 3 (R) appears to be an attractive precursor for the cyclization from
the positions 3 and 4 opening the route to natural products in
this series [22].

°Since we have synthesized 3-trimethylsilyl indole from
o-tolyl isonitrile [20] we have investigated the silylation of
isonitriles and related compounds.

The most significant results were obtained from PhN=C=S
and PhN=CCl$_2$ which led, upon silylation by the lithium/THF route
[23]. to silylated enediamines.

$$\equiv Si-N-C(Si\equiv) - C(Si\equiv)-N-Si\equiv \quad (\equiv Si = Me_3Si, HMe_2Si)$$
$$\quad\quad | \quad\quad\quad\quad\quad\quad\quad\quad | $$
$$\quad\quad Ph \quad\quad\quad\quad\quad\quad\quad Ph$$

The crystal structure of (I) ($\equiv Si=Me_3Si$) exhibited a twisted
double bond :

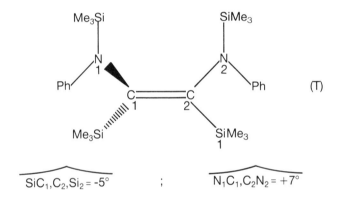

The C=C bond length was normal (1,350 Å) whereas the C-N
bond was longer than the normal (1,468 Å instead of 1,42 Å).

These examples as well as many models synthesized by us
or in other laboratories confirm the wide potentiality of the
polysilylation reaction to elaborate sterically hindered organosilicon models, structures having no   equivalent in
pure organic chemistry or precursors of polyfunctional substrates. Moreover all the X-ray structures analyses confirmed
the assignments established from NMR (^{1}H, ^{13}C, ^{29}Si) data which
appears once more to be a sure technique for structure de-
terminations.

1. R. Calas and J. Dunoguès, J.Organometal.Chem. Library, 1976, 2, 277.
2. M. Laguerre, J. Dunoguès and R. Calas, Tetrahedron Lett., 1981, 22, 1227.
3. J. Dunoguès, D. N'Gabe, M. Laguerre, N. Duffaut and R. Calas, Organometallics, 1982, 1, 1525.
4. M. Laguerre, D. N'Gabe, C. Biran et J. Dunoguès, Tetrahedron submitted for publication.
5. M. Laguerre, J. Dunoguès and R. Calas, Tetrahedron, 1978, 34, 1823.
6. J. Dunoguès, R. Calas and N. Ardoin, J.Organometal.Chem., 1972, 43, 127.
7. M. Laguerre, J.Dunoguès and R. Calas, Tetrahedron Lett., 1980, 21, 831.
8. G. Félix, J. Dunoguès, F. Pisciotti and R. Calas, Angew. Chem., Intern.Ed.Engl., 1977, 16, 488.
9. G. Félix, J. Dunoguès and R. Calas, Angew.Chem., Intern.Ed. Engl., 1979, 18, 409.
10. G. Félix, J. Dunoguès and R. Calas, J.Chem.Res., 1980, 236.
11. D. N'Gabe and J. Dunoguès, to be published.
12. K. Shiina and H. Gilman, J.Amer.Chem.Soc., 1966, 88, 5367. See also D.H. Ballard, T. Brennan, F.W.G. Fearon, I. Haiduc, K. Shiina and H. Gilman, Pure Appl.Chem., 1969, 19, 449 and ref. therein.
13. B. Bennetau, D. N'Gabe and J. Dunoguès, to be published.
14. M. Laguerre, G. Félix, B. Rezzonico and J. Dunoguès, to be published.
15. a) M. Laguerre, G. Félix, J. Dunoguès and R. Calas, J.Org. Chem., 1979, 44, 4275.
    b) M. Laguerre, G. Félix, R. Calas and J.Dunoguès, J.Org. Chem., 1982, 47, 1423.
16. C. Biran, B. Efendene and J. Dunoguès, J.Organometal.Chem., 1983, 253, C13.
17. S. Brownstein, J.Dunoguès, D. Lindsay and K.U.Ingold, J.Amer. Chem.Soc., 1977, 99, 2073.
18. A. Ekouya, J. Dunoguès, C. Biran, N. Duffaut et R. Calas, J.Organometal.Chem., 1979, 177, 137.
19. J-P. Picard, A. El Yusufi and J. Dunoguès, unpublished results.
20. B. Efendene, Dr.-Ing. Thesis, Bordeaux, 1983.
21. A.G.M. Barrett, D. Dauzonne and D.J. Williams, J.Chem.Soc. Chem.Comm., 1982, 636.
22. H.G. Floss, Tetrahedron, 1976, 32, 873.
23. C. Biran, B. Efendene, M. Laguerre and J. Dunoguès, to be published.

CHAPTER 16

NOVEL METAL-MEDIATED CYCLOADDITION REACTIONS OF 1,2-DISILACYCLOBUTENES WITH DIENES

C. S. Liu*, C. H. Lin and C. Y. Lee Department of Chemistry,
National Tsing Hua University, Hsinchu, Taiwan, ROC

It is well established that under uv irradiation $Fe(CO)_5$ catalyzed hydrosilylation of olefins proceeds via either an olefin complex intermediate followed by oxidative addition of $HSiR_3$, or a hydrido-silyl complex which then takes up the olefin on further dissociation of carbonyls [1-4]. Both cases result in the formation of $(olefin)(H)Fe(SiR_3)(CO)_3$, which is considered to be the catalytic species but has not been isolated.

In the case of the hydrosilylation of conjugated dienes, typical reactions involve the formation of 1,4 addition to the dienes _via_ a η^3-allyl intermediacy. For example [5-7],

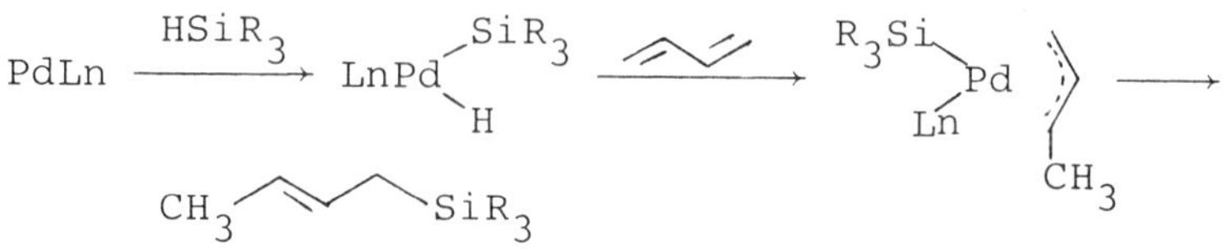

$$PdLn \xrightarrow{HSiR_3} LnPd\diagdown \begin{smallmatrix} SiR_3 \\ H \end{smallmatrix} \quad \diagup\diagup \longrightarrow \begin{smallmatrix} R_3Si \\ \diagdown \\ Ln \end{smallmatrix} Pd \diagup\diagdown \begin{smallmatrix} \\ \\ CH_3 \end{smallmatrix} \longrightarrow$$

$$CH_3\diagup\diagdown\diagup\diagdown SiR_3$$

Closely related is the cycloaddition reactions of conjugated dienes with disilacyclobutene, the latter is a special class of vinyldisilanes that has often been considered as the more stable form of 1,4-disilabutadiene [8-12]. For example, the cycloaddition between 1,1,2,2-tetramethyl-1,2-disilacyclobutene and acetylenes has been interpreted by the following reaction scheme:

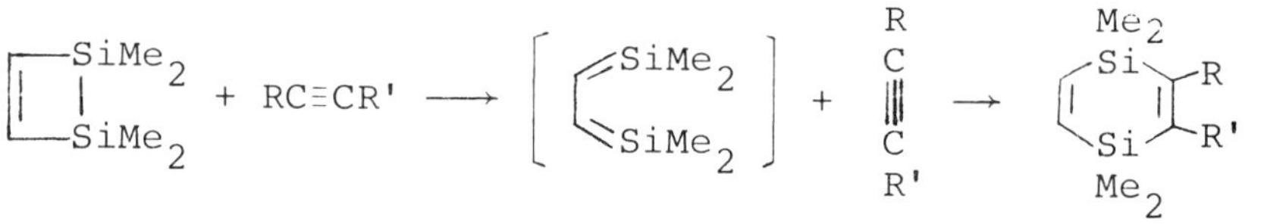

$$\square\begin{smallmatrix} SiMe_2 \\ SiMe_2 \end{smallmatrix} + RC\equiv CR' \longrightarrow \left[\begin{smallmatrix} SiMe_2 \\ SiMe_2 \end{smallmatrix}\right] + \begin{smallmatrix} R \\ C \\ \| \\ C \\ R' \end{smallmatrix} \longrightarrow \begin{smallmatrix} Me_2 \\ Si \diagdown R \\ Si \diagdown R' \\ Me_2 \end{smallmatrix}$$

Attempts to stablize silicon-carbon double bonded intermediates by transition metal carbonyls under both thermal and photochemical conditions have so far been unsuccessful [13]. Experiments aimed for stablizing disilabutadienes by transition metal carbonyls merely resulted in the formation of disila-metallocycles [14,15], the structure of some of these compounds have been determined by single crystal X-ray diffraction experiment [16]. We choose as a model reaction the cycloaddition between 1,3-butadiene system and 1,1,2,2-tetrafluoro-1,2-disila-cyclobutene with the hope that, should disilabuta-diene exist at all, the reactions may proceed via certain intermediates which would reveal the existence of such species.

Cycloaddition reaction of 1,3-butadiene with 3-tert-butyl-1,1,2,2-tetrafluoro-1,2-disilacyclobutene, compound 1, proceeded smoothly at temperatures above 100°C. The only type of products found was the product from 1,4 addition:

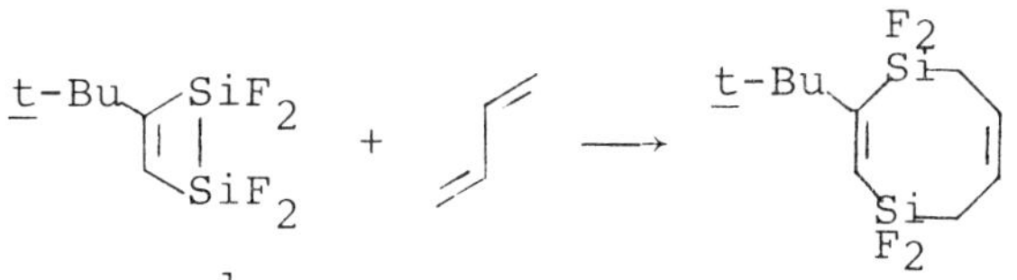

1 2

When the reaction was carried out under photochemical conditions and in the persence of iron pentacarbonyl, completely different types of products were obtained:

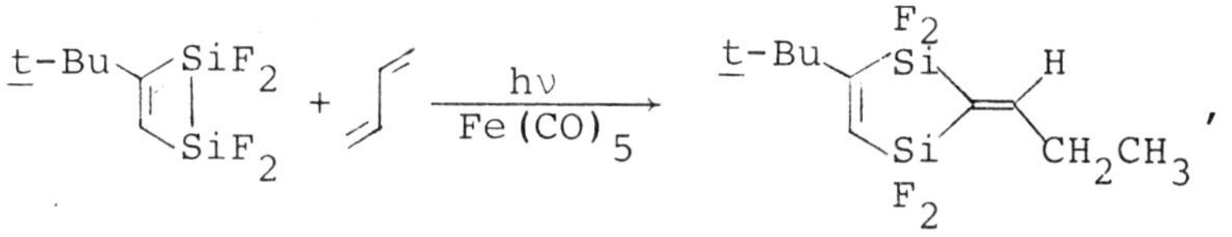

1 3a

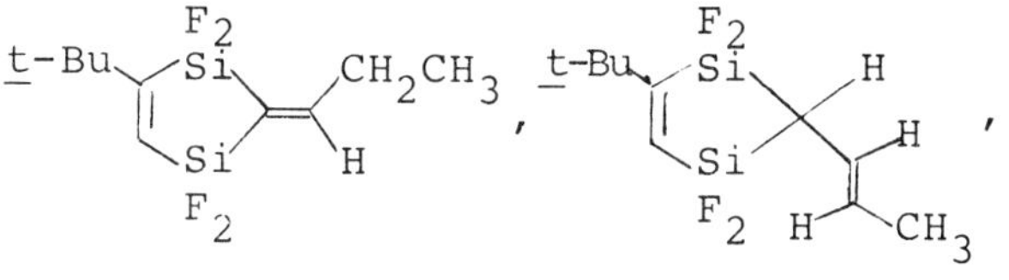

3b 4a

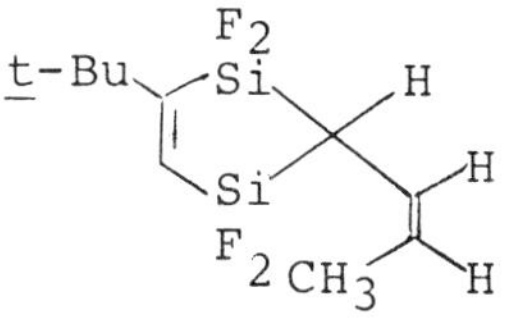

4b

This reaction proves to be photo-induced catalytical. When the reactions proceeded under uv irradiation at -30°C, the following reaction scheme

can be established with each intermediate isolated
and characterized experimentally:

The isolated compound 7 thermally decomposed
to products 3 and 4.

At $-10°C$, 7 was converted to another intermediate 8 slowly before the formation of final products
3 and 4. Intermediate 8 was proven to have a η^3-
allyl moiety by its ^1H and ^{13}C nmr spectra.

8

A plausible reaction mechanism can be proposed
for the reaction of 1,3-butadiene:

7 → 8 → 4a,4b ; 3a,3b

2,3-Dimethylbuta-1,3-diene reacted with 1
similarly and the products were found to be exclusively 9a and 9b. No products corresponding to 4a
and 4b in the reaction with butadiene were observed.
It seems that a methyl group on carbon 2 would favor
the migration of the second hydrogen from carbon 1.

1 + (CH₃ diene) $\xrightarrow[\text{Fe (CO)}_5]{h\nu}$ 9a

9b

If this is true, according to the mechanism described above, one may expect the reaction with isoprene would lead to products 10a, 10b, 11a, 11b, and 12 because attacks on both carbon 1 and carbon 4 are possible in the case.

Indeed, when isoprene was reacted with 1 under the same experimental conditions, the expected five products were obtained exactly. The cycloaddition of 1 to these three butadienes apparently proceeds via a very unusual 1,1-addition.

When tungsten hexacarbonyl was used, the low temperature photochemical reaction yielded the same type of H-migration products as those in the reaction using iron pentacarbonyl, however, the thermal reaction gave unexpected 13a and 13b.

Compounds 13a and 13b were isolated and fully characterized by mass spectrometry and ^1H, ^{19}F and ^{13}C nmr spectroscopy. The characteristic quartet

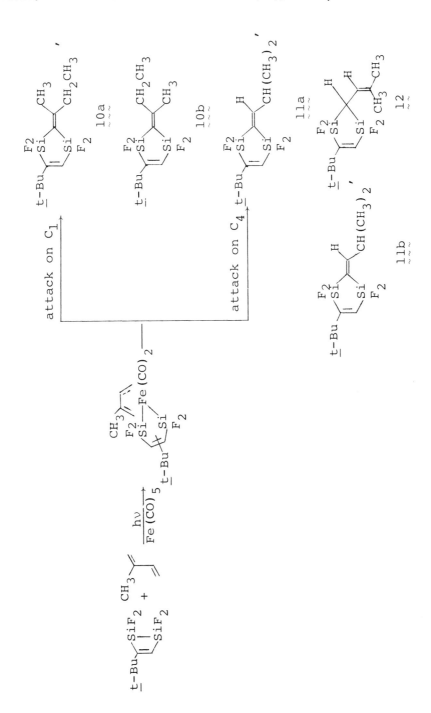

and doublet in the ^{19}F nmr spectrum for SiF and SiF$_3$
groups and the doublet (due to F coupling) of
triplets in the off-resonance ^{13}C spectrum for the
CH$_2$ carbons seem to be the unequivocal assignments of
the structures of 13. The position of the <u>tert</u>-butyl
group is determined on basis of the H-F coupling
constant in the SiF$_3$ resonance, as well as the
coupling pattern in the spectra of the olefinic carbons ($>$Si-C=C-SiF$_3$) in the off-resonance ^{13}C nmr
spectrum.

These results indicate a very unusual fluorine
migration assisted by the metal from one silicon to
the other. That the fluorine migration proceeds
alongside with the hydrogen migration may suggest a
common intermediate which allows both pathways to
proceed in a parallel manner:

Intermediate species 14, 14a and 14b involving
η^3-allyl and η^3-silallyl moieties are not expected
to be stable under the experimental conditions used
[2], however, in the case of iron pentacarbonyl, we
did observe the η^3-allyl intermediate, namely,
compound 8.

Compound 8 can be viewed as the stable form of
14a. When 8 was warmed up to 70°C, it converted to
3 and 4 quantitatively.

The isolation of 8 is a strong evidence that
the common intermediates, such as 14, do exist in
the cycloaddition reaction between butadiene and
disilacyclobutene, which allows both hydride migration and fluoride migration to take place.

When one considers the fact that hydride and
fluoride are the simple softest and hardest base
respectively, it follows quite naturally that one
would like to know the effect of the hardness (or
softness) of the metals on the reaction pathways.

The results of the reactions using $Cr(CO)_6$,
$Mo(CO)_6$ and $W(CO)_6$ as catalyst are listed in Table 1.
These reactions were carried out under the same
experimental conditions (0°C photochemically). The
softer tungsten assists hydride migration to form 4,
whereas the harder chromium facilitates fluoride
migration only. In the middle case of molebdenum,
products via both H-migration and F-migration are
obtained.

Table 1 – Products of the cycloaddition reactions* mediated by various metal carbonyls.

metal carbonyl	product	
$Cr(CO)_6$		
$Mo(CO)_6$		
$W(CO)_6$		

* photochemically at 0°C in n-pentane solutions

Similarly, cycloaddition reaction of 1,3-cyclohexadiene with 1 proceeds smoothly at 100°C, and the only product was from 1,4 addition.

When the reaction was carried out in the presence of $Fe(CO)_5$ photochemically, compouds 16a, 16b and 17 were obtained.

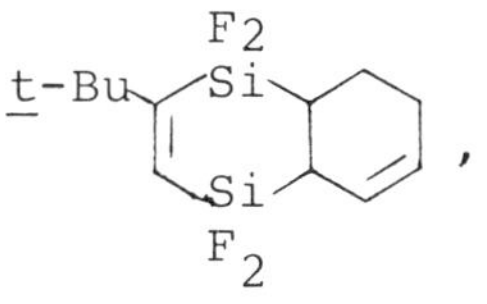

$$
\text{t-Bu} \underset{\underset{F_2}{Si}}{\overset{\overset{F_2}{Si}}{}} \text{(16b)}, \quad \text{t-Bu} \underset{\underset{F_2}{Si}}{\overset{\overset{F_2}{Si}}{}} \text{(17)}
$$

16b 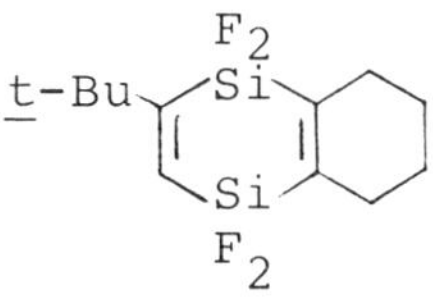   17

A reaction intermediate 18 was isolated and fully characterized by mass spectrometry, ^{1}H, ^{19}F and ^{13}C nmr spectroscopy in solutions, and x-ray diffraction in single crystal. At 100°C, 18 decomposed to products 16a, 16b and 17. During the process of thermal decomposition of 18, one more intermediate, 19, was isolated, which led to the formation of 17 at elevated temperature.

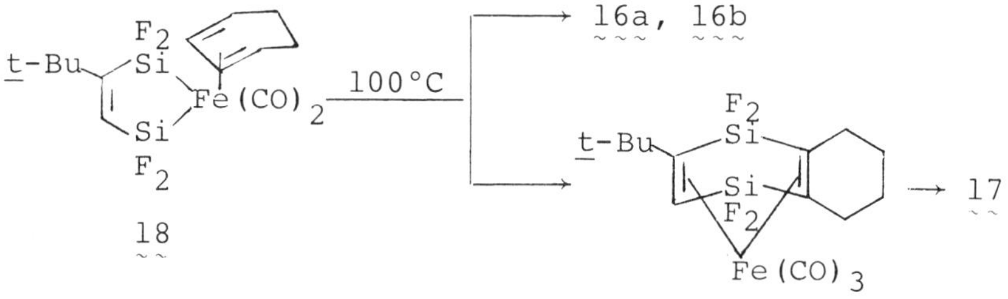

$$
18 \;\xrightarrow{\;100°C\;}\;
\begin{cases}
16a,\ 16b \\[4pt]
19 \;\longrightarrow\; 17
\end{cases}
$$

18

19

When the products 16a/16b, and 17 were treated with Fe(CO)$_5$ separately under the same reaction condition as that of the decomposition of 18, there is no observation of interconversion among the three. The only reaction appeared to occur was the reversed conversion of 17 to 19. In fact, pure 19 was conveniently obtained by this means.

The observation of 17 as one of the major products in the reaction may suggests that the initial attack of the silicon atom is on carbon 2 of cyclohexadiene. One plausible reaction mechanism can be proposed as follows:

It is interesting to note that when $W(CO)_6$ was used instead of $Fe(CO)_5$, only one product, 16b, was obtained. The intermediate in the reaction of $W(CO)_6$ was also isolated and characterized as 20.

The structures of 18 and 20 were determined by
single crystal x-ray diffraction experiments.   In
the case of 18, the Fe-disilacycle 5-membered ring
is puckered in such a way that two silicon atoms are
located within 2.80Å~3.20Å to C_1 and C_2.   On the
other hand, the structure of 20 shows that the W-
disilacycle ring is flat and oriented nearly perpendicular to the cyclohexadiene ring so that
only one silicon is at the vincinity (~3.0Å) of C_1
and C_2 whereas the other silicon (near t-Bu group)
locates very far away (>5Å) from any of the four
diene carbons.

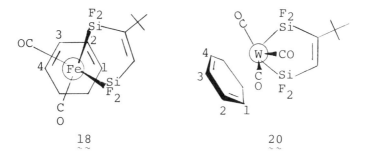

The fact that 20 led to only one isomer, 16b,
is a strong evidence that the initial attack of Si
atom (the silicon atom away from the t-Bu group) was
on carbon 2 instead of carbon 1.   Otherwise one
would expect to obtain 16a instead of 16b.   This
evidence is in agreement with the reaction mechanism
proposed above.

Acknowledgement.   The authors thank the Chinese
National Research Council for financial support to
this work.   One of us (CHL) thanks the Nuclear Energy
Research Institute for a research fellowship.

REFERENCES

1. Mitchener, J. C.; Wrighton, M. S. Photogenera-
    tion of Very Active Homogeneous Catalysts Using
    Laser Light Excitation of Iron Carbonyl Pre-
    cursors. J. Amer. Chem. Soc., 1981, 103, 975-
    977.

2. Schroeder, M. A.; Wrighton, M. S. Pentacar-
    bonyliron(o) Photocatalyzed Reactions of Tri-
    alkylsilanes with Alkenes. J. Organomet. Chem.,
    1977, 128, 345-358.

3. Schroeder, M. A.; Wrighton, M. S. Pentacar-
    bonyliron(o) Photocatalyzed Hydrogenation and
    Isomerization of Olefins. J. Amer. Chem. Soc.,
    1976, 98, 551-558.

4. Jetz, W.; Graham, A. G. Silicon-Transition
    Metal Chemistry. I. Photochemical Preparation
    of Silyl(transition metal) Hydrides. Inorg.
    Chem., 1971, 10, 4-9.

5. Langova, J.; Hetflejs, J. Catalysis by Metal
    Complexes. XXV. Hydrosilylation of 1,3-Butadiene
    Catalyzed by Palladium(II) Complexes. Collect.
    Czech. Chem. Commun., 1975, 40, 420-431.

6. Langova, J.; Hetflejs, J. Catalysis by Metal
    Complexes. XXVI. Palladium-catalyzed Hydro-
    silylation of 1,3-Butadiene by Trimethylsilane.
    Collect. Czech. Chem. Commun., 1975, 40, 432-441.

7. Sakurai, H.; Kamiyama, Y.; Nakadaira, Y.
    Chemistry of Organosilicon Compounds. 82. Palla-
    dium Complex-catalyzed Reactions of Hexaorgano-
    disilanes with Dienes. Chem. Lett., 1975, 887-
    890.

8. Barton, T. J.; Kilgour, T. A. An Alternative
    Mechanism for the Formation of 1,4-Disilacyclo-
    hexa-2,5-dienes from Acetylenes and Silylenes.

J. Amer. Chem. Soc., 1976, 98, 7746-7750.

9. Ishikawa, M.; Kumada, M. Photochemically
    Generated Silicon-Carbon Double Bonded Inter-
    mediates. J. Organomet. Chem., 1978, 149,
    37-48.

10. Ishikawa, M. T.; Sugaya, T. and Kumada, M.
    Photolysis of Organopolysilanes. A. Novel
    Addition Reaction of Aryl Substituted Disilanes
    to Olefins. J. Amer. Chem. Soc., 1975, 97,
    5923-5923.

11. Ishikawa, M.; Kumada, M. Photochemically
    Generated Silicon-Carbon Double-Bonded Inter-
    mediates VIII. Photolysis of Phenyldisilane
    Derivatives in the Presence of Olefins. J.
    Organomet. Chem., 1978, 162, 223-238.

12. Sakurai, H.; Nakadaira, Y. New Photochemical
    Reaction of Vinyldisilanes Through Silaethene
    or Silacyclopropane Intermediates. J. Amer.
    Chem. Soc., 1976, 98, 7424-7425.

13. Liu, C. S. Trapping the Carbon-Silicon Doubly
    Bonded Intermediates from Vinyldisilanes.
    Proceeding NSC (ROC), 1982, 6, 279-287.

14. Liu, C. S.; Cheng, C. W. Cycloaddition of
    Disilacyclobutenes in the Presence of Nickel
    Tetracarbonyl. J. Amer. Chem. Soc., 1975,
    97, 6746-6749.

15. Chi, Y.; Liu, C. S. Photochemical Preparation
    of Transition-Metal Carbonyl Compounds with
    1,1,2,2-Tetrafluoro-1,2-disilacyclobutenes as
    Ligands. Inorg. Chem., 1981, 20, 3456-3460.

16. Hseu, T. H.; Chi, Y.; Liu, C. S. Crystal and
    Molecular Structure of (1,1,4,4-Tetrafluoro-2-
    tert-butyl-1,4-disilabut-2-ene) Molebdenum (II)
    Pentacarbonyl. Inorg. Chem., 1981, 20, 199-204.

CHAPTER 17

CATALYSIS OF HYDROSILYLATION AND METATHESIS OF VINYL-SUBSTITUTED SILANES

Bogdan Marciniec Faculty of Chemistry, A. Mickiewicz University, 60–780 Poznan, Poland

Hydrosilylation of vinylsilanes and vinylsiloxanes has recently been one of the most studied processes involving the addition of Si-H to C=C bonds. The reaction can be regarded as a simplified model for an activated cure of the polydimethylsiloxane chain with unsaturated groups by polyfunctional silicon hydrides. The reaction of hydrosilanes (hydrosiloxanes) with vinylsilanes (vinylsiloxanes) occurs essentially according to the following scheme:

$$\equiv SiH + CH_2=CHSi\equiv \begin{cases} \longrightarrow \equiv SiCH(CH_3)Si\equiv & \alpha\text{-adduct} \\ \\ \longrightarrow \equiv SiCH_2CH_2Si\equiv & \beta\text{-adduct} \end{cases} \qquad (1)$$

Regioselectivity of the examined reaction is a result of the substituent effect at silicon both in vinyl and in hydro-silanes (siloxanes) as well as of catalyst used. In the presence of some Pt [1], Os [1], Fe [2,3] and Ru [4] complexes also products of the dehydrogenative double hydrosilylation occur according to the general equation:

$$\equiv SiH + 2CH_2=CHSi\equiv \longrightarrow \equiv SiCH=CHSi\equiv + CH_3CH_2Si\equiv \qquad (2)$$

HYDROSILYLATION OF VINYLTRI(CHLORO)(METHYL)SILANES
CATALYZED BY PALLADIUM PHOSPHINE COMPLEXES

In the reaction of chloro(methyl)substituted silanes, mostly β-adduct is formed when chloroplatinic acid in isopropanol (Speier catalyst) [5] or in cyclohexanone [4] as well as Wilkinson complex $RhCl(PPh_3)_3$ [4] and heterogenized Pt-complexes [6] are used as catalysts. Complexes of nickel [4], as well as recently reported Speier catalyst [7], lead to the formation of a mixture of α- and β-adducts.

Our study on catalysis of hydrosilylation of vinyltrichlorosilane by trichlorosilane in the presence of palladium(0) phosphine complex showed the selective formation of α-adduct —1,1-bis(trichlorosilyl)ethane [8]. This unusual regioselectivity of the reaction became a subject of more detailed examination including a substituent effect at silicon and various Pd(0) and Pd(II) precursors used [9]. The $PdCl_2(PPh_3)_2$, which can be formed when a mixture of palladium, triphenylphosphine and trichlorosilane is heated to 120°C [10] as well as when $Pd(PPh_3)_4$ is treated with trichlorosilane [11], appeared to be the real selective catalyst for the 1,1-bis(trichlorosilyl)ethane preparation (9). However, in the absence of phosphine (Pd, $PdCl_2$) no reaction occurs at all. Replacement of even one chlorosubstituent by methyl one both in hydro- and/or in vinylsilane causes a change in regioselectivity and leads to the selective formation of β-adduct. Besides, a series of studies of the hydrosilylation of vinyltri(chloro)methyl-silanes by trichloro(methyl)silanes in the presence of $Pd(PPh_3)_4$ catalyst revealed the following order of reactivity of hydrosilanes $Cl_3SiH > Cl_2MeSiH \gg Cl(Me_2)SiH$ while the yield of hydrosilylation of $CH_2=CHSiMe_nCl_{3-n}$ (n=0-3) by trichlorosilane is constant and equal to 90% of the adduct [9].

The present study allows us to propose a general scheme of the reaction examined in the presence of the initial Pd(0) and Pd(II) phosphine complexes (Scheme 1).

Mechanistic considerations have to include the oxidative addition of hydrosilanes to get an intermediate (1) formed as we know, from Pd(0) and/or Pd(II) phosphine complexes. A coordination of vinylsilane before the oxidative addition cannot be excluded. Electron-withdrawing chloro substituents at silicon stabilize the complex (1). The introduction of methyl substituent at silicon instead of chloro one stabilizes the Pd-Si bond and we thus

Scheme 1

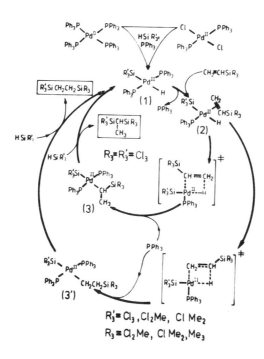

observe a marked decrease in the hydrosilylation
yield. Different regioselectivity of vinyltrichlorosilane by trichlorosilane seems to be a result of a
nucleophilic hydride attack in the complex (2) on
β-carbon of vinylsilane(according to the proposed
transition state). Such a mechanistic path can be
supported by the observed regioselectivity, in the
addition of trichlorosilane to other vinyl compounds
with electron-withdrawing functional groups, e.g.
acrylonitrile and styrene, which leads in the presence of palladium phosphine complexes also to α-addu-
cts [11].

COMPETITIVE HYDROSILYLATION AND METATHESIS OF VINYL-
SUBSTITUTED SILANES CATALYZED BY RUTHENIUM PHOSPHINE
COMPLEXES

Numerous transition metal complexes form adducts
with vinylsilanes [2,3,12,13]. Some of them, e.g.
$Fe(CO)_4 \cdot (CH_2=CHSiMe_3)$ catalyze the hydrosilylation
of vinyltrimethylsilanes by various chloro(methyl)hy-
drosilanes /70-80°C, 3 h, increased pressure/ which
gives, beside α- and β-adducts, also the products of
dehydrogenative hydrosilylation [2,3]. Hydrosilylation of vinyltrimethylsilane as well as of vinyltrialkoxysilanes by trialkoxysilanes in the presence of
chloroplatinic acid [4,14], Wilkinson catalyst [4]
heterogenized Pt-catalyst [15] as well as cobalt and
rhodium carbonyls [16,17] yields mostly β-adducts,
whereas α-adducts are observed as by-products. Our
previously reported study on hydrosilylation of vinyltrialkoxysilanes by trialkoxysilanes, catalyzed by
various ruthenium(II) and ruthenium(III) precursors,
showed the reaction to result in the unsaturated product $(RO)_3SiCH=CHSi(OR)_3$ beside β-adduct.
The content of the latter in the products increases
markedly when the reaction is carried out in an
excess of vinyltrisubstituted silanes [4,18]. Under
such conditions the reaction can occur due to the above
mentioned dehydrogenative hydrosilylation (eq. 2)
and/or the metathesis of vinyltrialkoxysilanes:

$$2 \ CH_2=CHSi\equiv \ \rightleftharpoons \ \equiv SiCH=CHSi\equiv + CH_2=CH_2 \qquad (3)$$

The latter process was found to be the first efficient example of metathesis of organosilicon compounds, occurring also in the absence of hydrosilanes [19].
The earlier reported study on the examined reaction has shown the following general features of the
catalysis of the hydrosilylation of C=C bond by
$RuCl_2(PPh_3)_3$ and $RuCl_3(PPh_3)_3$ precursors [18];
- predominance of trialkoxysubstituted silanes and
  vinylsilanes in the formation of ruthenium-silyl
  active intermediates,
- requirement of a low coordination number of the initial ruthenium intermediates active only in the absence of solvent,
- assumption of the $Ru^I \longrightarrow Ru^{III}$ oxidative addition
  step in hydrosilylation, which bases on the promo-
  ting role of molecular oxygen, complexation of
  ruthenium by vinylsilanes, isolation of intermediates and other experimental data.

In order to find the origin of the unsaturated product (UP), 1,2-bis(triethoxysilyl)ethene, observed in the reaction of triethoxysilane with vinyltriethoxysilane, the stoichiometry of the reaction was examined under standard conditions [20].

Table 1 — Concentrations of the main products and parent substances during the hydrosilylation of vinyltriethoxysilane by triethoxysilane and competitive metathesis of vinyltriethoxysilane catalyzed by $RuCl_2(PPh_3)_3$ (followed by GLC).

Reaction time [h]	$[CH_2=CHSi\equiv]$ [M]	$[\equiv SiH]$ [M]	[trans-UP] [M]	$[\beta\text{-adduct}]$ [M]	$[EtSi\equiv]$ [M]
0	4.20	—	—	—	—
0.5[a]	4.05	0.42	0.07	0.00	0.00
2.5	1.67	0.02	1.04	0.34	0.03
5	0.85	0.02	1.48	0.34	0.03

$110^{\circ}C$, $[cat] = 0.83 \cdot 10^{-3}M$, air; [a]hydrosilane introduced

The results presented in Table 1 provide convincing evidence that hydrosilylation is accompanied by metathesis of vinyltriethoxysilane. The latter is considerably enhanced in the presence of hydrosilanes. Since the hydrosilylation reaction is stopped in the presence of any solvent, kinetic measurements were carried out in 10- and 20-fold excess of vinylsilane. The pseudo first-order observed rate constants k_{obs}, the rate of hydrosilane uptake $V_0(SiH)$ and the rate of β-adduct formation, $V_0(\beta)$, can be regarded as tests for the hydrosilylation. On the other hand, the rate of the unsaturated product formation, $V_0(UP)$, is practically a measure of the metathesis occurring under the same conditions. All the kinetic data show the first order with respect to hydrosilane although a slight downward deviations are observed on the plot of $V_0(SiH)$ vs. $[SiH]_0$ Moreover, the second order of the reaction with respect to vinylsilane was determined. Selected data on the dependence of the reaction rate on the catalyst concentration are shown in Fig.1 [20].

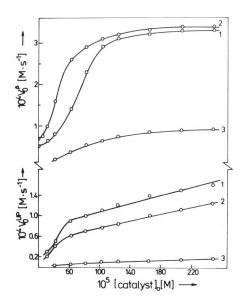

Figure 1 — The plot of the $V_0(\beta)$ and $V_0(UP)$ vs. the catalyst concentration for the reaction of triethoxysilane with vinyltriethoxysilane catalyzed by $RuCl_3(PPh_3)_3$ [1], $RuCl_2(PPh_3)_3$ [2,3] in air [1,2] and in argon [3], $t_{activ.}$ = 0.5h, temp. 110°C, [SiH]$_0$ = 0.42 M, [CH$_2$=CHSi]$_0$ = 4.2 M.

The complex course of the curves indicates the decreasing reaction order with respect to the initial catalyst for the hydrosilylation, essentially from 1⟶0, although under oxygen-promoting action in the range of small concentrations of the precursor, the second order can be evaluated. The first-order dependence is observed for the metathesis reaction. Compilation of the activation energy (E_a) of the competitive reaction, determined from the plots: $\ln k_{obs}$ (SiH) vs. 1/T and $\ln V_0$ (β) [V_0 (UP)] vs. 1/T, are given in Table 2 [20]. Considering the above E_a values, the direct conclusion can be drawn that metathesis of vinyltriethoxysilane occurs via entropy-controlled stages, i.e. via reaction pathways different than hydrosilylation, in spite of the fact that the active intermediates in metathesis can also catalyze the

Table 2. Activation energies for the hydrosilylation $E_a(SiH)$, $E_a(\beta)$ and metathesis $E_a(UP)$ of vinyltriethoxysilane catalyzed by phosphine complexes of ruthenium.

Precursor	$t_{activ.}$	Gas atmosph.	$E_a(SiH)$ $\left[\frac{kcal}{mole}\right]$	$E_a(\beta)$ $\left[\frac{kcal}{mole}\right]$	$E_a(UP)$ $\left[\frac{kcal}{mole}\right]$
$RuCl_2(PPH_3)_3$	0.5	air	9.2 ± 0.5	11.4 ± 0.6	4.7 ± 0.3
	3	air	9.8 ± 0.6	12.1 ± 0.5	3.8 ± 0.3
	0.5	argon	13.6 ± 0.6	–	4.7 ± 0.3
$RuCl_3(PPh_3)_3$	0.5	air	12.2 ± 0.5	13.6 ± 0.8	3.0 ± 0.5

$[cat] = 0.83 \cdot 10^{-3}$ M, $[SiH]_o = 0.42$ M, $[SiCH=CH_2]_o = 4.2$M

hydrosilylation, what results from different values of $E_a(SiH)$ and $E_a(\beta)$.

^{a}Hydrosilylation of vinyltriethoxysilane, contrary to that of alkenes[20], also occurs in the absence of dioxygen, though its presence causes several-fold enhancement in the rate of hydrosilylation as well as of metathesis (see Fig.1). Moreover, it is interesting to note a marked increase in the UP/β ratio under the promoting role of oxygen. Apparently, the activation of the catalyst by dioxygen gives rise to the formation of more active intermediates, above all for metathesis but showing also activity for hydrosilylation. Photoinduced catalysis by the ruthenium precursors in the oxygen atmosphere leads to increase in metathesis rate, $V_0(UP)$, as well as in hydrosilylation rate $V_0(\beta)$ (Fig. 2) [20].
It can be explained by the generation of high concentrations of metathesis active species (e.g. metal-carbenes [21]) or of coordinatively unsaturated centers via phosphine ligand direct dissociation or its oxidation to phosphine oxides [22].
All kinetic data [20] as well as the consideration of the reaction mechanisms on other metal centers [3] and of hydrosilylation of olefins with other electron-withdrawing groups [23] allow to present the following scheme for the hydrosilylation of vinyltrialkoxysilanes by hydrosilane which is changed a little in comparison with the one given previously [18]. (Scheme 2).

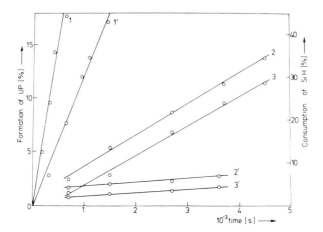

Figure 2 — Conversion of SiH (1,1') and formation of 1,2-bis(triethoxysilyl)ethene (UP) (2,2', 3,3') vs. time of the reaction of triethoxysilane with vinyltriethoxysilane occurring in the presence of $RuCl_2(PPh_3)_3$ in air (1–3) – near-UV irradiated, λ = 290–450 nm (1'–3') – no irradiation 110°C; (for 1,1' 102°C); no solvent, $[SiH]_0 : [SiCH=CH_2]_0$ = 1:10 (1,1',2,2') and 1:100/3,3'/; [cat] = 0.83·10^{-3} M.

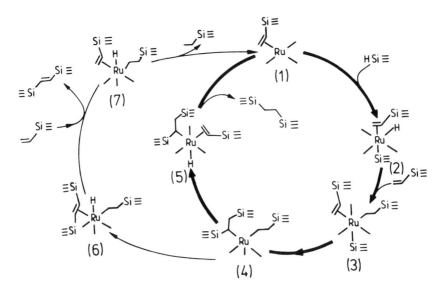

Scheme 2 — Proposed mechanism of hydrosilylation of vinyltrialkoxysilanes.

The mechanism postulates the existence of vinyl-silane-ruthenium intermediate which is stabilized by electron-withdrawing alkoxy groups at silicon. The catalytic cycle involves the oxidative addition of hydrosilane followed by the next vinylsilane molecule coordination what results in δ-ethylsilyl interme-diate (3) which does not undergo reductive elimina-tion because of the strong π-coordination of the vi-nylsilane. The silicon shift from the metal to the latter molecule is proposed to form σ-complex (4). Next two pathways are possible. The first of them is β-hydride shift from β-ethylsilyl ligand which pro-ceeds through the hydride (5) to β-adduct. However the abstraction of hydrogen from the α-silyl-β-ethylsilyl ligand can give the unsaturated product (UP) coordinated to the ruthenium center (6) followed by the olefin exchange at this center. Reductive eli-mination of the β-ethylsilylruthenium hydride (7)gi-ves the ethylsilane and regenerates the initial acti-ve species (1). The latter path is apparently much less plausible under conditions examined. Formation of 1.2-bis(triethoxysilyl)ethene mainly a result of metathesis occurring competitively with hydrosilyla-tion. One of the catalyst systems for the metathesis involves metal carbene formation by interaction of the parent olefin compound with the transition metal center. In the reaction examined, an initiation of the ruthenium carbenes may proceed according to the following equation [21]:

$$
\begin{array}{ccccc}
\mathrm{CH_2{=}CHSi\equiv} & & \mathrm{CH{=}CHSi\equiv} & & \mathrm{CH{-}CH_2Si\equiv} \\
| & \longrightarrow & | & \longrightarrow & \| \\
[\mathrm{Ru}] & & [\mathrm{Ru}]{-}\mathrm{H} & & [\mathrm{Ru}]
\end{array} \quad (4)
$$

The determination of Ru-H bond ($\nu=1965$ cm^{-1}) in the isolated intermediate obtained by treatment of $\mathrm{RuCl_2(PPh_3)_3}$ with vinyltriethoxysilane can adduce in support of such a reaction sequence.

In the presence of hydrosilane, an initiation of metal carbene can occur via the δ-complex (see Sche-me 2).

$$
\begin{array}{ccccc}
\mathrm{CH_2CH_2Si\equiv} & & \mathrm{CHCH_2Si\equiv} & \xrightarrow{\mathrm{HSi}\equiv} & \mathrm{CHCH_2Si\equiv} \\
| & \longrightarrow & \| & \xleftarrow{} & \| \\
\equiv\mathrm{Si\text{-}[Ru]} & & \equiv\mathrm{Si\text{-}[Ru]\text{-}H} & & [\mathrm{Ru}]
\end{array} \quad (5)
$$

In excess of vinylsilane the other metal carbe-nes can be generated according to the simplified mo-del given by Basset et al. [24] based on stereoche-mistry considerations:

$$
\begin{array}{c}
\equiv SiCH_2 \quad H \\
\diagdown \diagup \\
C \\
\parallel \quad\quad CH_2 \\
Ru\cdots\parallel \\
CH \\
\diagup \\
\equiv Si
\end{array}
\rightleftharpoons
\begin{array}{c}
\equiv SiCH_2CH\!\!=\!\!CH_2 \\
\vdots \\
Ru\!\!=\!\!CHSi\equiv
\end{array}
\tag{6}
$$

It is obvious that metal carbenes can also initiate the hydrosilylation cycles.

METATHESIS OF VINYL-SUBSTITUTED SILANES CATALYZED BY
MCl_3–COCATALYST SYSTEM (M= Ru, Rh)

Our experimental work on metathesis of vinyl-substituted silanes has allowed to find catalytic systems based on ruthenium and rhodium chlorides with some Lewis acids such as trialkylsilanes, trialkoxy-silanes, tetraalkoxysilanes, $LiAlH_4$, $NaBH_4$ as cocatalysts effective of metathesis of vinyltrialkoxysilanes, vinylalkyldialkoxysilanes as well as a number of other vinylsilanes and vinylcyclosiloxanes [25]. Selective data are given in Table 3 for vinylmethyl-diethoxysilane. In both cases a mixture of trans- and cis- product is formed.

Table 3 — Effect of cocatalyst on the yield [%] of the metathesis product (followed by GLC) occurring in the presence of rhodium(III) and ruthenium(III) chloride catalysts.

$$
/EtO/_2MeSiCH\!\!=\!\!CH_2 \longrightarrow
\begin{array}{c}
H \quad\quad Si\equiv \\
\diagdown \diagup \\
CH\!\!=\!\!CH \\
\diagup \diagdown \\
\equiv Si \quad\quad H
\end{array}
+
\begin{array}{c}
H \quad\quad H \\
\diagdown \diagup \\
CH\!\!=\!\!CH \\
\diagup \diagdown \\
\equiv Si \quad\quad Si\equiv
\end{array}
$$

trans–UP   cis–UP

Cocatalyst	$RhCl_3$		$RuCl_3$	
	trans–UP	cis–UP	trans–UP	cis–UP
–	4	trace	trace	–
$(EtO)_3SiH$	77	8	66	3
Et_3SiH	75	13	60	2
$LiAlH_4$	70	4	43	trace
$NaBH_4$	8	trace	61	trace
Reaction conditions:	$130^{\circ}C$, 0.5 h		$130^{\circ}C$, 3 h	
	air, 4 mmoles of vinylsilane, 0.04mmole catalyst and cocatalysts, no solvent			

These catalytic systems were also succesfully used in polymetathesis of divinyldisiloxanes giving oligomers of the general formula [25].

$$CH_2=CH-[\overset{|}{\underset{|}{Si}}-O-\overset{|}{\underset{|}{Si}}-CH=CH]_n-\overset{|}{\underset{|}{Si}}-O-\overset{|}{\underset{|}{Si}}-CH=CH_2 \qquad n=1-7$$

The new products were isolated, identified by spectroscopic methods, elemental analysis as well as, in the case of difunctional siloxanes, by gel chromatography. The possible scheme for the metathesis of vinyl-substituted silanes catalyzed by ruthenium and rhodium compounds is given below (Scheme 3).

Scheme 3

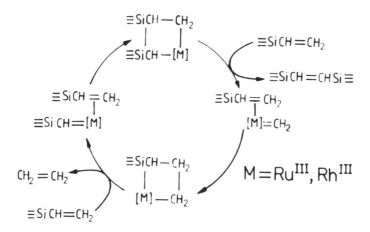

The metal carbene mechanism which is the most plausible working hypothesis (and now the accepted one) in the concept of metathesis, was first given by Herisson and Chauvin [26] and for Wilkinson catalyst by Cardin et al. [27]. The idea is based on the metal carben-metallocyclobutane intermediates formation in the catalytic cycle.

The obtained results prove that this reaction can be applied as a novel catalytic method for synthesis of unsaturated monomers and oligomers as well as for modification of polymers.

ACKNOWLEDGEMENTS

The material discussed in this paper encompases the research and unpublished results of several co-workers of mine, notably J.Guliński, W.Urbaniak, L.Rzejak and E.Maćkowska whose contributions are gratefully acknowledged.

REFERENCES

1. J.L.Speier, Adv.Organomet.Chem., 1979, 17, 407.
2. G.A.Nurtdinova, G.A.Tailunas, V.P.Yur´ev, Izv. Akad.Nauk SSSR, Ser.Khim., 1981, 2652.
3. G.A.Tailunas, G.V.Nurtdinova, V.P.Yur´ev, G.A.Tolstikov, S.R.Rafikov, Izv.Akad.Nauk SSSR, Ser.Khim., 1982, 914.
4. B.Marciniec, J.Guliński, W.Urbaniak, Pol.J.Chem., 1982, 56, 287.
5. V.M.Vdovin, A.D.Petrov, Zh.Obshch.Khim., 1960, 30, 838.
6. Ger.Pat., 2245187.
7. V.D.Sheludyakov, V.I.Zhun, S.D.Vlasenko, V.N.Bochkarev, T.F.Slyusarenko, A.V.Kisin, V.M.Nosova, G.N.Turkel´taub, E.A.Chernyshev, Zh.Obshch.Khim., 1981, 51, 2022.
8. B.Marciniec, J.Guliński, W.Urbaniak, Synth.React. Inorg.Metal-Org.Chem., 1982, 12, 139.
9. B.Marciniec, E.Maćkowska, J.Guliński, W.Urbaniak, Z.anorg.allg.Chem., (submitted for publication).
10. P.Fitton, J.E.Mc Keon, Chem.Commun., 1968, 4.
11. J.Tsuji, M.Hara, K.Ohno, Tetrahedron, 1974, 30 2143.
12. J.W.Fitch, K.C.Chan, J.A.Froelich, J.Organomet. Chem., 1978, 160, 477.
13. J.W.Fitch, W.T.Osterloh, J.Organomet.Chem., 1981, 213, 493.
14. C.L.Frye, W.T.Collins, J.Org.Chem., 1970, 35, 2694.
15. Ger.Pat., 2400039.
16. G.K.I.Magomedov, O.V.Shkol´nik, B.A.Izmailov, S.A.Sigachev, Koord.Khim., 1980, 6, 761.
17. G.K.I.Magomedov, O.V.Shkol´nik, Zh.Obshch.Khim., 1980, 50, 1103.
18. B.Marciniec, J.Guliński, J.Organomet.Chem., 1983, 253, 349.
19. B.Marciniec, J.Guliński, J.Organomet.Chem., 1984, 266, C19.
20. B.Marciniec, J.Guliński, unpublished results.
21. K.J.Ivin, Olefin Metathesis, Academic Press, 1983.
22. R.A.Faltynek, Inorg.Chem., 1981, 20, 1357.
23. I.Ojima, T.Fuchikami, M.Yatabe, J.Organomet.Chem., 1984, 260, 335.
24. J.M.Basset, M.Leconte, Fundamental Research in Homogeneous Catalysis, 1979, 3, 285.
25. B.Marciniec, L.Rzejak, unpublished results.
26. J.L.Herisson, Y.Chauvin, Macromol.Chem., 1970, 141, 161.
27. D.J.Cardin, M.J.Doyle, M.F.Lappert, J.Chem.Soc., Chem.Commun., 1972, 927.

CHAPTER 18

STEREOCHEMICAL CONTROL IN ORGANIC SYNTHESIS USING SILICON

Ian Fleming University Chemical Laboratory, Lensfield Road, Cambridge, England

CONTROLLING THE POSITION OF A DOUBLE BOND

One of our earliest pieces of work in the organosilicon field was to show how a silyl group, placed on the appropriate atom, controlled the outcome of a cationic cyclisation (Scheme 1)[1]. In 1974-5, when this work was being done, the allylsilane (**2**) was easy to make using the well-established Wurtz-like reaction of the allyl chloride (**1**) and trimethylsilyl chloride,

Scheme 1

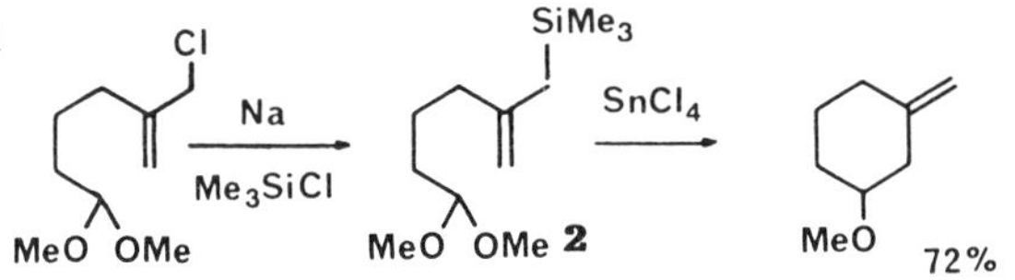

but the other possible silanes (**3** and **4**) were not accessible at that time. Thus we were early exposed to a major limitation in the idea that a properly placed silyl group could control the outcome of a carbonium ion reaction. The limitation was that there were, in common use, only two ways of introducing a silyl group into organic molecules, electrophilic silylation of an organometallic carbon nucleophile and hydrosilation of an alkene or alkyne, and these were not powerful enough easily to solve the problems inherent in synthesising molecules like **3** or **4**, let alone more complicated and highly functionalised silanes. We were led therefore to develop our silyl-cuprate reagent, which would be a nucleophilic silyl reagent. Our starting point was Gilman's phenyldimethylsilyl-lithium; this reagent combined with a copper(I) salt to give a reagent able to react (Scheme 2) with terminal acetylenes, with tertiary allylic acetates, and with αβ-unsaturated carbonyl compounds[2]. We have also shown [3] that the same reactions occur with a reagent derived from trimethylsilyl-lithium, even though this is less easy to make and incorporates HMPA in the solvent system.

Scheme 2

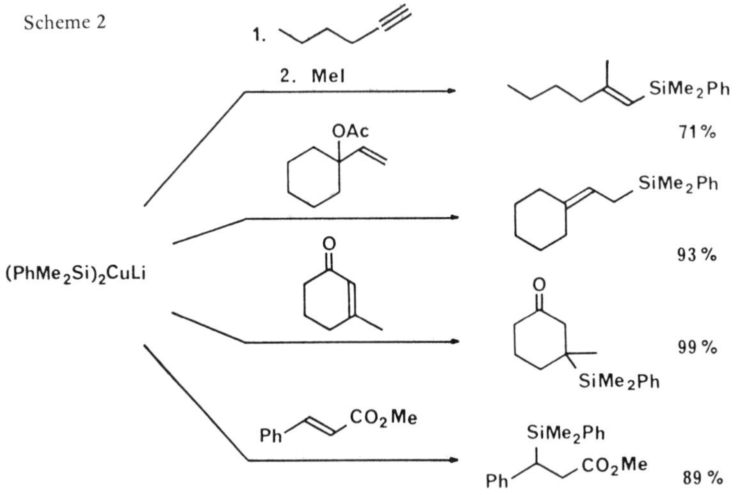

These reagents have allowed us to solve the problem which our earlier work set us: how to make the allylsilane (3) and the vinylsilane (4). These syntheses are shown in Schemes 3 and 4, the former being a specific example of a general method [4] for

Scheme 3

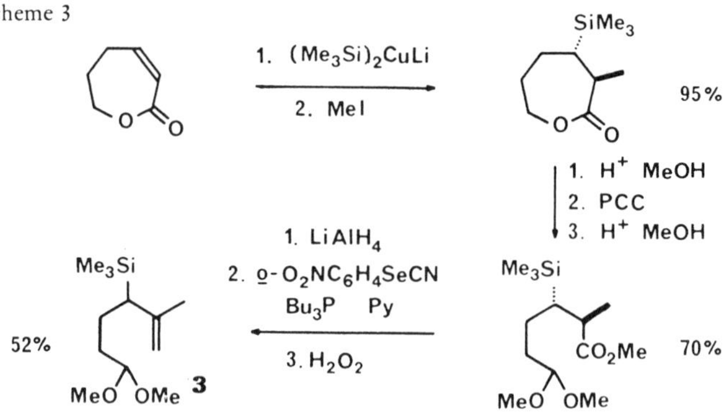

Scheme 4

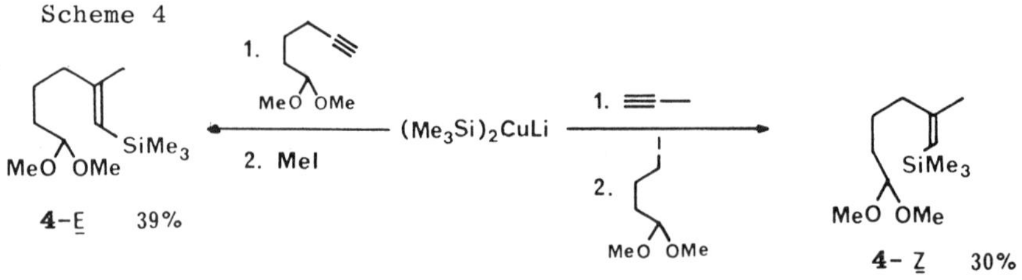

the synthesis of allylsilanes with the silicon at the more-sub-
stituted end of the allyl fragment.   In both cases, the silanes
cyclised exclusively to the appropriate alkene.   We have, there-

fore, demonstrated the regiospecificity of the reaction, and the
complete control that silicon can exert.   The contrast is most
vividly made by comparing these results with the situation where
silicon is absent: all three alkenes are produced[5].

CONTROLLING STEREOCHEMISTRY

The work I have just described brought one of our lines to
fruition.   Our newer work uses a silyl group quite differently.
The purpose now is to control stereochemistry rather than the
position of a double bond.   One branch of our present work is to
investigate the stereochemistry of electrophilic attack on allylsilanes.   This is now well-established to be *anti* stereospecific,
both by our own work[6], and by the work of Wetter, of Eschenmoser, of Kitching, and most particularly, of Kumada[7].   I do
not want to deal with this topic here, except to observe that
the same principle carries over from the allylsilane (5) to

the β-silylenolate (6).
The starting point for our work was the observation that conjugate attack by the silyl-cuprate reagent on methyl cinnamate,
followed by methylation of the intermediate enolate (7, Scheme
5), gave very largely one diastereoisomer of the alkylated ester
(8).   In this case, we have now found that the geometry of the
double bond makes no difference to the diastereoselectivity:
conjugate addition, protonation, and treatment with LDA gave the
geometrical isomer (9) of the enolate, and methylation of this
enolate gave the same diastereoisomer (8) with the same select-

ivity as before. We also found that conjugate addition, followed by protonation of the intermediate enolate (10) gave largely
(85:15) the other diastereoisomer (11). We published this work

Scheme 5

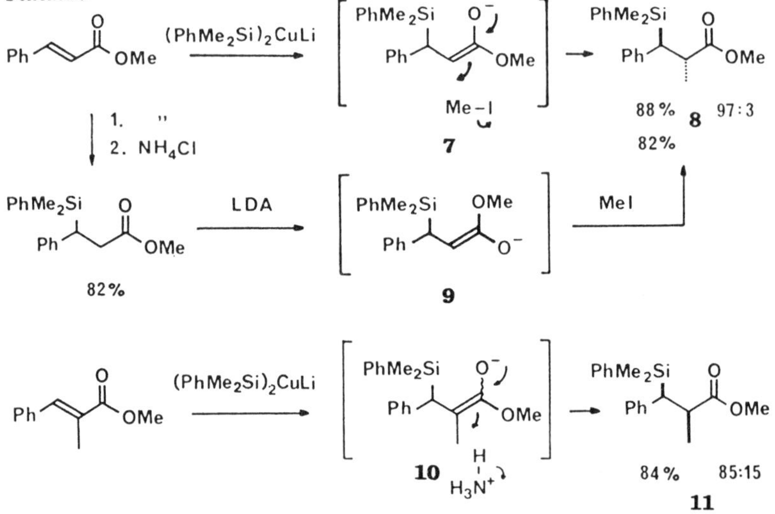

earlier this year, together with the proof of stereochemistry
of the products[8]. The stereochemistry is accounted for by
analogy with the stereochemistry of attack on allylsilanes.
As with the allylsilane (5), the β-silylenolate has a strongly
preferred conformation (6), in which the hydrogen on the chiral
centre is the only group small enough to eclipse the double bond,
leaving the phenyl and silyl groups staggered about the hydrogen
atom on the double bond. Electrophilic attack then takes place
on the side away from the silicon, either for electronic or
steric reasons, or for both combined. This pattern of diastereo-

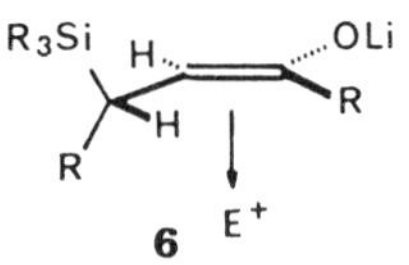

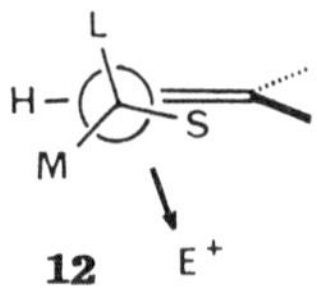

selectivity is also consistent with Houk's prediction[9]. He and
his coworkers suggest that the reactive conformation for electrophilic attack on a trigonal carbon adjacent to a chiral centre
is 12, which is closely similar to 5 and 6, if we assume that
the silyl group is the large group.
    With this pattern of results in hand, we were faced with a
large task. There are many places (Scheme 6, arrows) where
variations can be made, and we wanted to know how changes at
each of them affected the diastereoselectivity.

Scheme 6

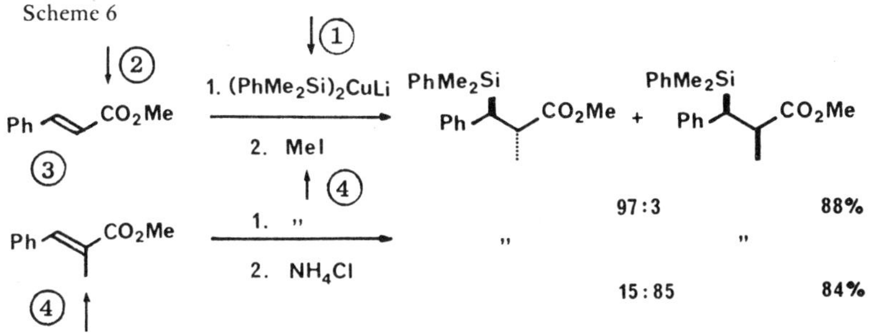

Variation at the silyl group (arrow 1)

Let us begin with the silyl group; how important is it to have a metal here? We have looked at compounds with just alkyl and phenyl groups, and there is diastereoselectivity (Scheme 7) The sense of the diastereoselectivity is consistent with Houk's picture (12), but the diastereoselectivity is not of the high order which we find in the silicon series. On the other hand,

Scheme 7

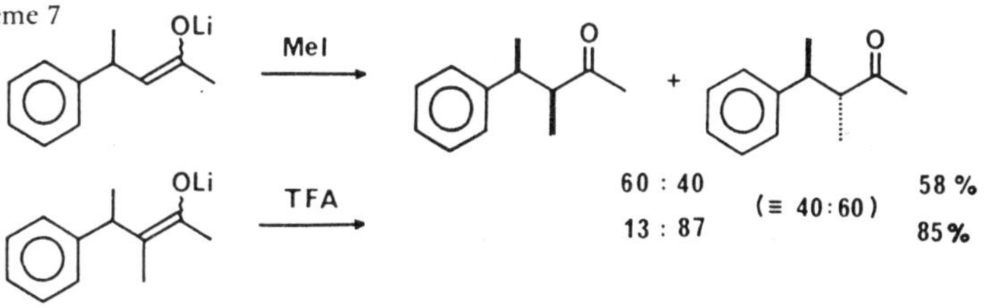

a trimethyltin group (Scheme 8) does induce high diastereoselectivity. This result gave us some stereodefined tin-containing compounds, which we have used, so far, to investigate the

Scheme 8

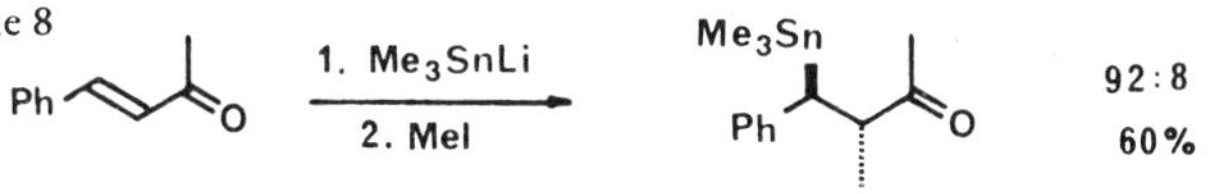

Scheme 9

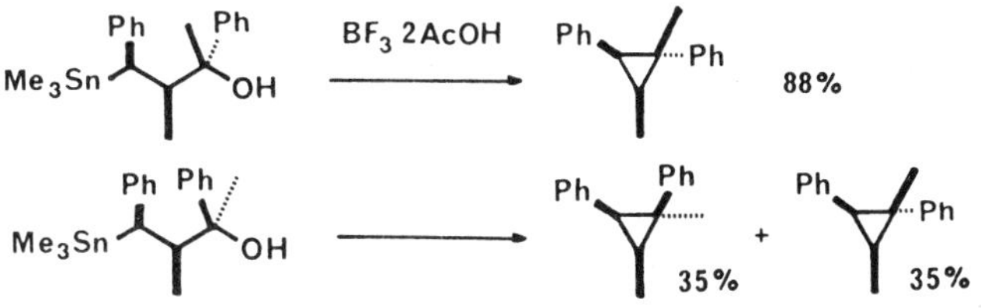

Scheme 10

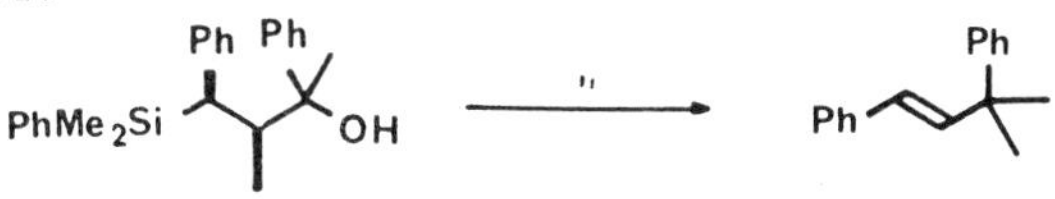

stereochemistry of a cyclopropane-forming reaction (Scheme 9),
in which we found complete *inversion* of configuration at the nucleophilic carbon, and a· residue of *inversion* even at the tertiary benzylic carbon[10].   Incidentally, this reaction contrasts
sharply with the corresponding reaction in the silicon series
(Scheme 10), where *rearrangement* is the major pathway.

Variation in the type of enolate (arrow 2)

Esters are not the only source of enolates for our reaction;
the results with other carbonyl compounds are shown in Scheme 11.
The reaction is highly diastereoselective, and always in the

Scheme 11

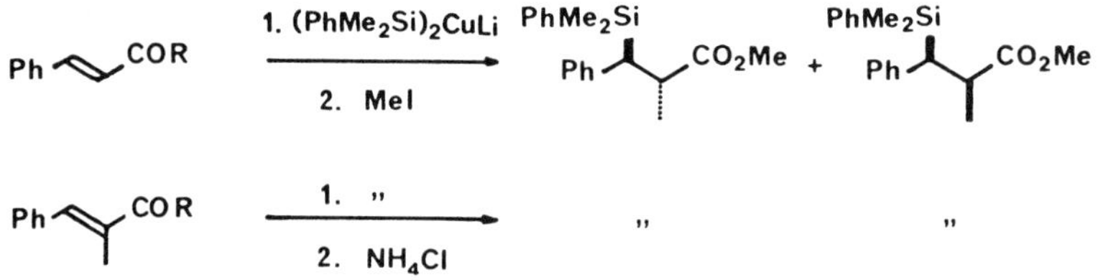

R	Methylation		Protonation		Proof
OMe	97:3	88 %	15:85	84 %	✓
Me	98:2	57%	30:70		✓
H	92:8	74%	11:89	86%	✓
Ph	high	70%	—	0%	✓
NMe$_2$	97:3	86%	18:82	83%	✓
CN for COR	54:46	65%	14:86	77%	✓

same sense, except in the case of the nitrile. This is easily
explained (Scheme 12): the nitrile anion has no substituent
eclipsing the C-C bond attached to the nucleophilic trigonal
carbon, with the result that the conformation (**13**) will no longer
be much preferred over the alternatives. This same idea also

Scheme 12

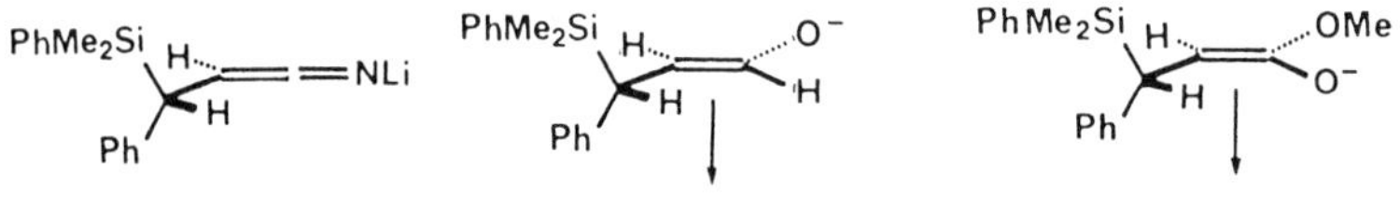

explains why the aldehyde enolate (**14**) is somewhat less diastereoselective than the other enolates: the same C-C bond is
syn periplanar with a C-H bond, and again the conformational
preference will not be quite as strong as it is in the enolate
(**15**) derived from the ester. The stereochemistry of the enolate
double bond in **14** is trans, as we could see from the NMR spectrum
of the silyl enol ether derived from it[8].

Variation of the group attached to the chiral centre (arrow 3)

Here we found that alkyl groups induced diastereoselectivity in the same sense as the phenyl group had, but somewhat less effectively (Scheme 13): the larger the alkyl group, the lower the diastereoselectivity in the alkylation reaction. Here we

Scheme 13

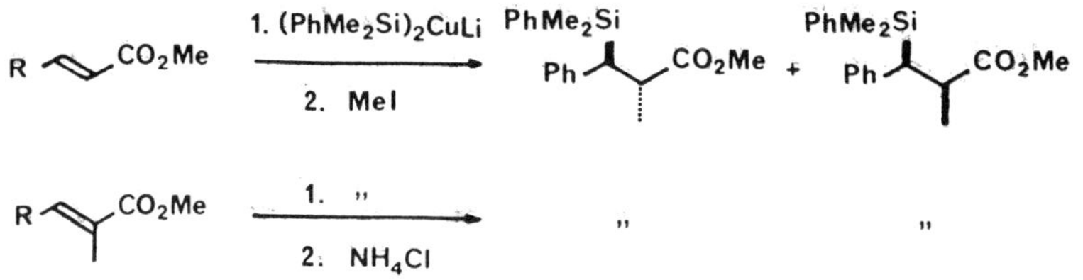

R	Methylation		Protonation		Proof
Ph	97:3	88 %	15:85	84 %	✓
Me	91:9	78 %	13:87	82 %	
Pr^i	85:15	95%	4:96	56%	✓
Bu^t	66:34	83%	4:96	38%	✓

come across a problem in measuring the effective size of the various groups, something which we need to know if we are to find out whether the diastereoselectivity is steric or electronic in origin. The A values[11](Scheme 14) suggest that silyl is larger than isopropyl, but smaller than phenyl. On the other

Scheme 14

	A value	r*
Me$_3$Si	2.5	2.1
Me$_3$Sn	1.1	
Ph	3.1	1.6
Me	1.7	1.8
Pr^i	2.1	2.2
Bu^t	~5	3.6

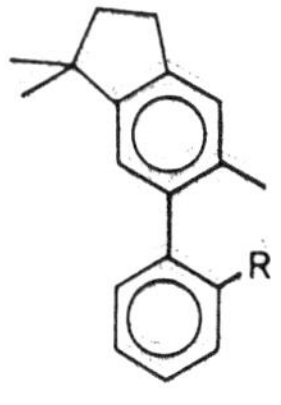

16

hand, the "effective radius" r*, which is measured[12] by the ease with which the R group can squeeze past a methyl group in the biphenyl (16), indicates that a silyl group is effectively smaller than isopropyl, and that a phenyl group is effectively smaller than a methyl. Both sets of numbers indicate that a t-butyl group is larger than the other groups. We conclude, tentatively, that the diastereoselectivity is largely determined by electronic factors, but that these are offset by steric effects: as the alkyl group R gets larger, the diastereoselectivity of the methylation reaction falls off, but does not change round. The diastereoselectivity in the protonation holds up with the larger alkyl groups, presumably because protonation is less sterically demanding. The high diastereoselectivity with the phenyl group may be explained in two ways. If the r* value is reliable, phenyl is effectively smaller than methyl, and the diastereoselectivity is therefore even higher for phenyl than for methyl. Or, if the r* value is misleading and the phenyl group is larger than the methyl, it is electonically a relatively electron-withdrawing substituent, and hence contributes to increase the electronic disparity between the silyl and the R group. The disparity between the silyl and the methyl is less, and the diastereoselectivity is correspondingly less. Finally, the good diastereoselectivity with the tin compound (Scheme 8) is also consistent with a large electronic component: its A value suggests that the tin group is effectively smaller than a phenyl group, yet attack takes place *anti* to the tin group.

Variation of the electrophile (arrow 4)

There is considerable scope for varying the electrophile. In the first place we looked at other alkyl halides (Scheme 15). Again we get high diastereoselectivity in the alkylation reaction, but protonation of the enolates derived from the esters (**18**, R = $PhCH_2$ and R = Pr^i) gave, as major products, the same diastereoisomers as the benzylation and isopropylation of the enol-

Scheme 15

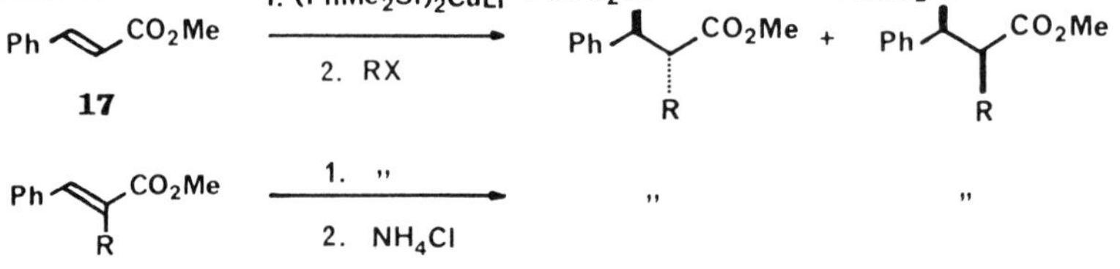

RX	Alkylation		Protonation	
MeI	97:3	88%	15:85	84%
EtI	95:5	83%	20:80	77%
Bu^nI	94:6	86%	27:73	77%
Pr^iI	95:5	26%	60:40	78%
$PhCH_2Br$	97:3	74%	71:29	66%
$CH_2{=}CHCH_2Br$	95:5	76%	31:69	83%
MeO_2CCH_2Br	98:2	50%	10:90	82%

ate derived from cinnamate (**17**). Our explanation for these two anomalies is that we are now looking at reactions of an enolate

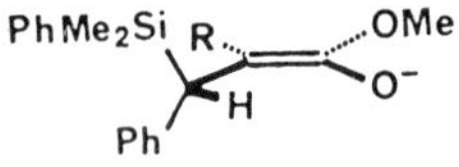

19

(**19**), in which the R group is larger than a methyl group. With a large R group in this position, the preferred conformation is no longer necessarily that shown in **19** (the silyl and phenyl groups will not easily accept a gauche interaction with a large R group). There is no simple argument for deducing what react- ive or preferred conformation will be adopted, and we are, at this stage, reduced, in the absence of calculations, to saying only that the direction of electrophilic attack becomes unpre- dictable.

The electrophiles used so far have been either a proton or tetrahedral carbon electrophiles. We have also examined trigon- al electrophiles, both those which react with the lithium enol- ates (**20**, Scheme 16) and those which react with the correspond- ing silyl enol ethers (**21**, Scheme 17). Again we notice that

Scheme 16

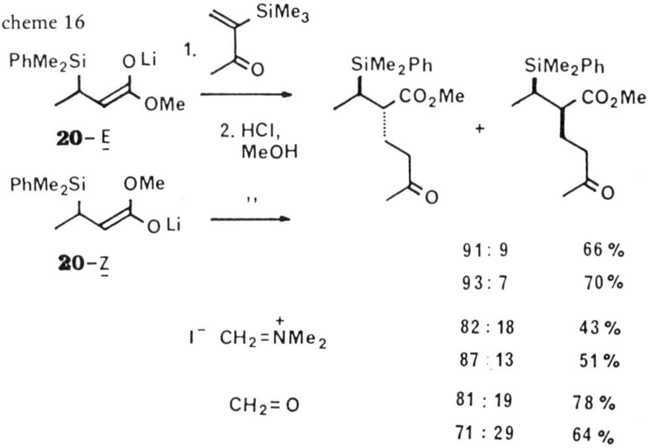

	91 : 9	66 %
	93 : 7	70 %
	82 : 18	43 %
	87 : 13	51 %
	81 : 19	78 %
	71 : 29	64 %

Scheme 17

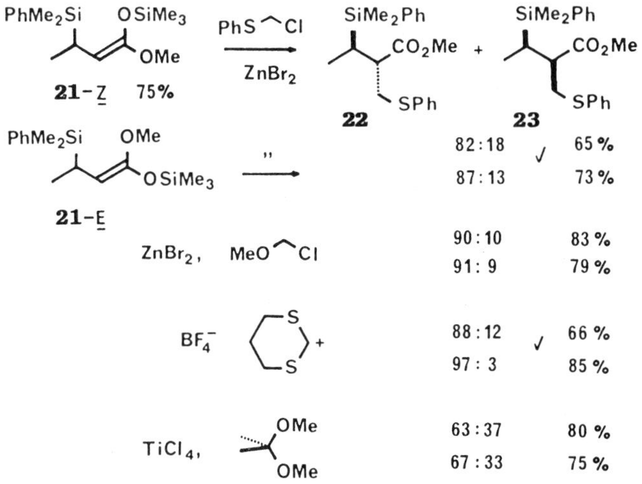

	82 : 18	65 %
	87 : 13	73 %
	90 : 10	83 %
	91 : 9	79 %
	88 : 12	66 %
	97 : 3	85 %
	63 : 37	80 %
	67 : 33	75 %

both geometrical isomers of **20** and **21** are available, just as they were in the cinnamate series (Scheme 5), and that both are attacked with moderate to high diastereoselectivity, but not always to quite the same degree. We have proved that the dia- stereoselectivity is in the same sense as before, by desulphur- isation of the esters (**22, 23**, and the product from the dithian- cation reaction), these gave the esters already known from the work shown in Scheme 13. Oxidative removal of the phenylthio group gave the unsaturated ester (**25**), which could also be made

Scheme 18

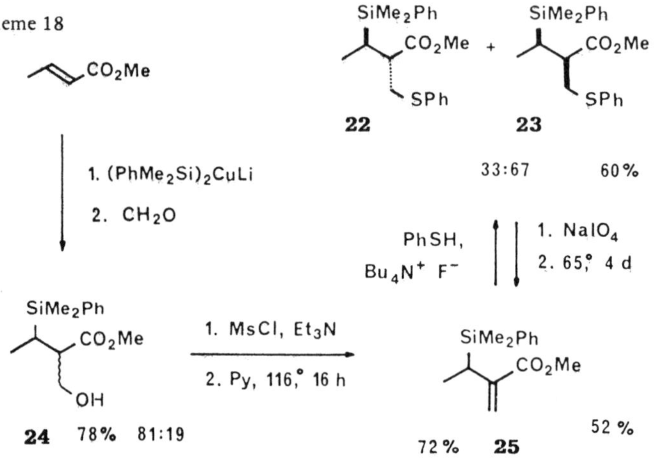

(Scheme 18) by dehydration of the 81:19 mixture of alcohols (**24**). Readdition of the phenylthio group now gave back the esters (**22** and **23**), but with the latter the major component, since the critical, stereochemistry-defining step is protonation.

The conversion of a silyl group to a hydroxyl

All these results gain in interest, when I reveal that *the phenyldimethylsilyl group can be converted in two steps, with retention of configuration, into a hydroxyl group*, as we have already published[13]. The first step is protodesilylation of the phenyl group, a well known reaction[14], and the second step is the oxidation of the silyl fluoride with a peracid in the presence of base, a reaction first investigated by Buncel and Davies, and revived independently by Kumada[15] and ourselves. Put together, these reactions enable us to convert our β-silyl esters into β-hydroxy esters, and the overall result (Scheme 19) is that we can synthesise β-hydroxy esters with high diastereo- selectivity. It was by using these reactions that we have been able to prove the configuration of the esters (**26**, R = H   and Me): the two steps (Scheme 20) gave us the known[16] hydroxy esters (**27**, R = H and Me).

Thus the work I have described is a new approach to the stereoselective synthesis of the products of an aldol reaction,

an area of much current interest among synthetic organic chemists. The wide range of variables, particularly of electrophile,
already establishes that this approach has something to add to

Scheme 19

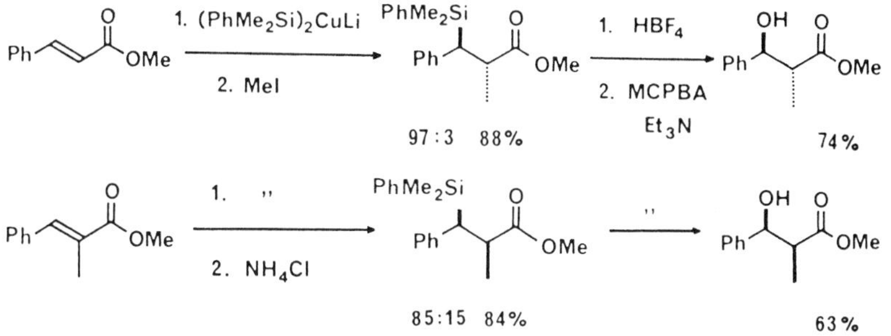

97:3 88%   74%

85:15 84%   63%

Scheme 20

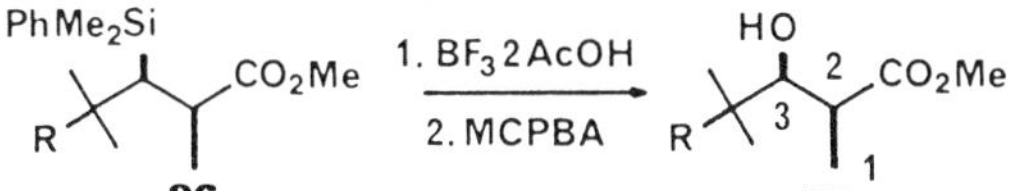

the existing methods. There is more potential yet, but I will
confine myself here to the diastereoselective synthesis of quaternary centres, which are not yet amenable to conventional aldol
methodology[17].

Quaternary centres

We can vary the electrophile in another sense: instead of
protonation, we can alkylate the enolate (e.g. **10**) derived from
an unsaturated ester which already has one alkyl group in place
(Scheme 21), and hence create a quaternary centre. This too is
highly diastereoselective, except when a large group (e.g. Pr^i)

Scheme 21

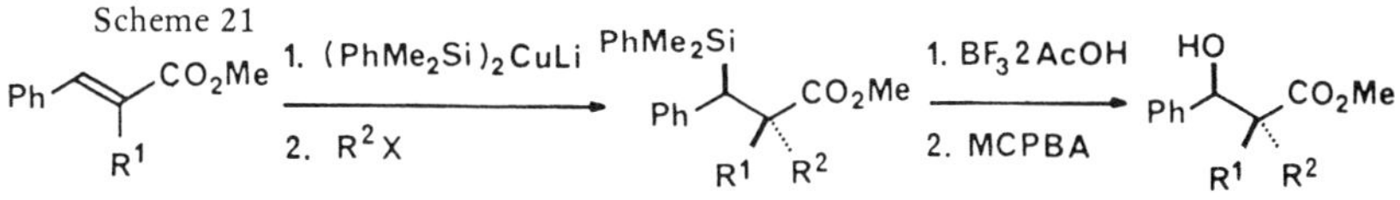

R^1	R^2X	Ratio	Yield	Yield
Me	EtI	83:17	94%	79%
Et	MeI	11:89	95%	74%
Me	$CH_2=CHCH_2Br$	83:17	90%	52%
$CH_2=CHCH_2$	MeI	20:80	90%	67%
Me	MeO_2CCH_2Br	92:8	45%	21%
MeO_2CCH_2	MeI	10:90	83%	
Me	Pr^iI	90:10	63%	
Pr^i	MeI	40:60	77%	

is in place: methylation is much less diastereoselective than it was with hydrogen in the α position (Scheme 5), although this time, in contrast to the protonation result (Scheme 15), it is not actually inverted. With one pair, we proved the configuration by reduction of the β-hydroxy esters (**28**) and acetal formation. The acetals (**29**) were susceptible to NOE-difference

Scheme 22

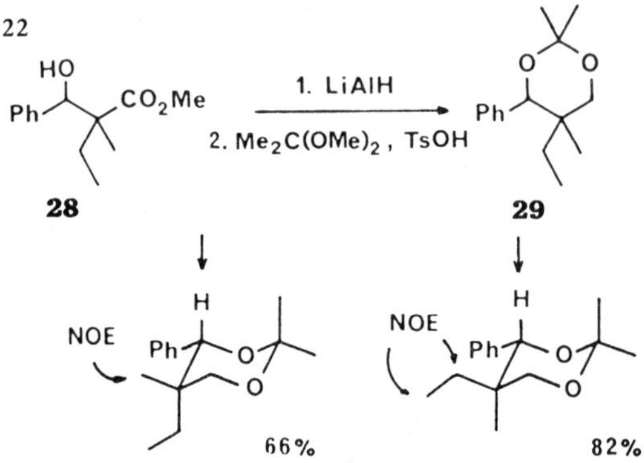

HO
Ph⟍⟍CO₂Me   1. LiAlH   ⟶   Ph⟍⟍   O   O
                2. Me₂C(OMe)₂, TsOH

28   **29**

                    ↓   ↓

        H   H
NOE Ph⟍⟍O⟍   NOE   ⟍Ph⟍⟍O⟍
          ⟍O   ⟍O

              66%   82%

experiments, which gave the results shown in Scheme 22.

CODA

   This concludes the main part of my lecture. I want to finish with two small points which this work has raised. We are all familiar with the cation-stabilising influence of a silylmethyl group. Should we expect that a silylmethyl group will destabilise an anion? Three pieces of evidence (Scheme 23) which support this expectation have come to our attention in the

Scheme 23

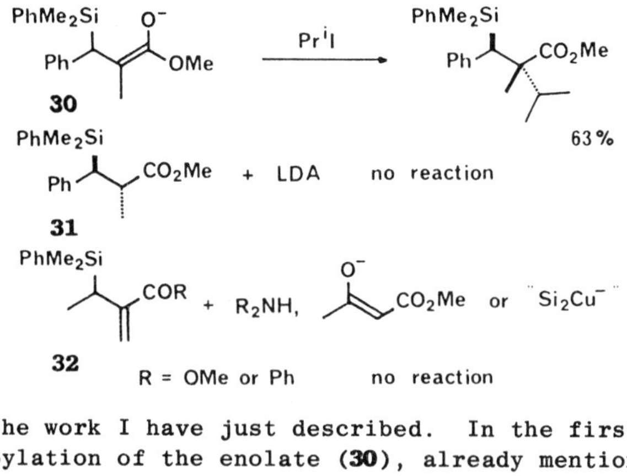

PhMe₂Si   O⁻   PhMe₂Si
Ph⟍⟍⟍OMe   PrⁱI   ⟶   Ph⟍⟍CO₂Me
30

                                                    63%

PhMe₂Si
Ph⟍⟍CO₂Me   + LDA   no reaction
31

PhMe₂Si
⟍⟍COR   + R₂NH,   O⁻⟍⟍CO₂Me   or ⟍Si₂Cu⁻
32
          R = OMe or Ph   no reaction

course of the work I have just described. In the first place, the isopropylation of the enolate (**30**), already mentioned in Scheme 21), is some indication that the enolate is extraordinary. Isopropylation of an isolated lithium enolate is not usually possible. In the second, we have found that standard, and even

quite forcing conditions have not allowed us to make enolates
from the ester (31). And in the third, we have not been able to
add several usually quite reactive nucleophiles, including our
silyl-cuprate reagent, to the ester (32); our only success has
been the addition of the phenylthio group (Scheme 18). With
this support for the idea that a silylmethyl group destabilises
an anion, we surmised that the ketone (33) would form an enolate

Scheme 24 :

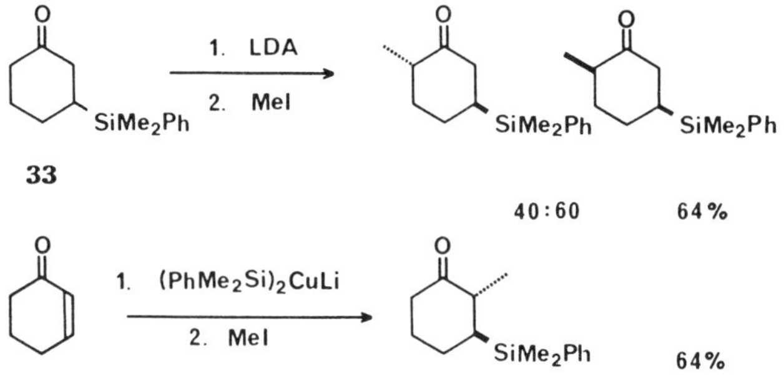

40:60   64%

64%

very largely on one side, away from the silyl group, and so it
proves to be (Scheme 24).

My second point picks up our lead that the diastereoselectivity is substantially electronic in origin. If this is the
case, it ought to be possible to pass chiral information along
a conjugated system. We have examined this possibility in one
case (Scheme 25). Conjugate addition of the silyl-cuprate reagent to E,Z-sorbate (34) and methylation of the enolate (35)
gave two diastereoisomers in a ratio of 60:40, with the diastereoisomer (36) as the major product, as shown by the conversion to the known[17] lactones (37). With chiral centres so far

Scheme 25

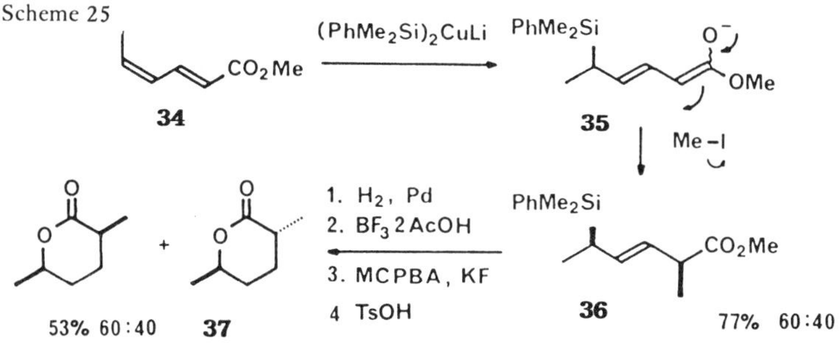

53% 60:40   37

77% 60:40

apart, it is most unlikely that any concentration of either diastereoisomer had taken place in this sequence. The diasetereo-
selectivity is low, but its direction is intriguing. The face
of the enolate double bond which has been attacked is on the
same side as the silyl group; in the extended conformation (38),
which is a redrawing of the enolate (35) using the bent-bond
model, we can see that the silicon-carbon bond is *anti* to the
upper bent bond adjacent to it, and this in its turn is *anti* to

the lower bent bond, which is actually attacked. Given the low diastereoselectivity with which chiral information and stereo-

38

specificity appear to be transmitted through double bonds[18], even a 60:40 ratio is interesting.

The unpublished work which I have described has been carried out in the last year by several members of my research group, to all of whom I am most grateful. They were: Hak-Fun Chow (Schemes 3 and 4), Jerry Lewis (Scheme 7), David Waterson (Schemes 11 and 21), John Hill (Schemes 13, 15, and 20), Jeremy Kilburn (Schemes 16, 17, and 18), David Parker (Scheme 21), Roger Smithers (Scheme 24), and Ken Takaki (Scheme 25).

REFERENCES

1. Fleming, I.; Pearce, A. J. Chem. Soc., Perkin Trans. 1, 1981, 251

2. Fleming, I.; Newton, T. W.; Roessler, F. J. Chem. Soc., Perkin Trans. 1, 1981, 2527. Fleming, I.; Marchi, D. Synthesis, 1981, 560. Ager, D. J.; Fleming, I.; Patel, S. K. J. Chem. Soc., Perkin Trans. 1, 1981, 2520.

3. Fleming, I.; Newton, T. W. J. Chem. Soc., Perkin Trans. 1, in press.

4. Fleming, I.; Waterson, D. J. Chem. Soc., Perkin Trans. 1, in press.

5. van der Gen, A.; Wiedhaup, K; Swoboda, J. J.; Dunathan, H. C.; Johnson, W. S. J. Am. Chem. Soc., 1973, 95, 2656.

6. Fleming, I.; Terrett, N. K. J. Organomet. Chem., 1984, 264, 99.

7. Wetter, H.; Scherer, P. Helv. Chim. Acta, 1983, 66, 118. Eschenmoser, A.; Jenkins, P. R.; Matassa, V. unpublished work. Wickham, G.; Kitching, W. J. Org. Chem., 1983, 48, 612. Hayashi, T.; Konishi, M.; Ito, H.; Kumada, M. J Am. Chem. Soc., 1982, 104, 4962. Hayashi, T; Ito, H; Kumada, M. Tetrahedron Lett., 1982, 23, 4605.

8. Bernhard, W.; Fleming, I.; Waterson, D. J. Chem. Soc., Chem. Commun., 1984, 28. Bernhard, W.; Fleming, I. J. Organomet. Chem., in press.

9. Paddon-Row, M. N.; Rondan, N. G.; Houk, K. N. J. Am. Chem. Soc., 1982, 104, 7162.

10. Fleming, I.; Urch, C. J. Tetrahedron Lett., 1983, 24, 4591.

11. Kitching, W.; Olszowy, H. A.; Drew, G. M.; Adcock, W. J. Org. Chem., 1982, 47, 5153. Kitching, W.; Doddrell, D.; Grutzner, J. B. J. Organomet. Chem., 1976, 107, C5. Eliel, E. L.; Allinger, S. J.; Morrison, G. A. Conformational Analysis, Wiley, New York, 1967.

12. Bott, G.; Field, L. D.; Sternhell, S. J. Amer. Chem. Soc.,

1980, 102, 5618.

13. Fleming, I.; Henning, R.; Plaut, H. J. Chem. Soc., Chem. Commun., 1984, 29.

14. Eaborn, C.; Bott, R. W.; Organometallic Compounds of the Group IV Elements, Vol. 1, Part 1, Macdiarmid, A. G., Ed., Marcel Dekker, New York, 1968, p. 410. See also, Kumada, M.; Ishikawa, M.; Maeda, S. J. Organomet. Chem., 1964, 2, 478.

15. Buncel, E.; Davies, A. G. J. Chem. Soc., 1958, 1550. Tamao, K.; Kakui, T.; Akita, M.; Iwahara, T.; Kanatani, R.; Yoshida, J.; Kumada, M. Tetrahedron, 1983, 39, 983.

16. Heathcock, C. H.; Pirrung, M. C.; Sohn, J. E. J. Org. Chem., 1979, 44, 4294, corrected by Heathcock, C. H., personal communication.

17. Pirkle, W. H.; Adams, P. E. J. Org. Chem., 1979, 44, 2169.

18. Eschenmoser, A., lecture at "Synthesis in Organic Chemistry", Cambridge, 1979.

CHAPTER 19

SELECTIVITY IN SILICON-MEDIATED ORGANIC REACTIONS: ORIGIN AND APPLICATION

R. Noyori, M. Hayashi and S. Hashimoto Department of Chemistry, Nagoya University, Chikusa, Nagoya 464, Japan

INTRODUCTION

Silicon has now become one of the most versatile elements in organic synthesis. Organic and inorganic silicon compounds display a multitude of functions in organic reactions, depending on the relative bond strength, relative electronegativity, and involvement of the valence 3p- or vacant 3d- atomic orbitals, participation of high-lying σ- or low lying σ^*-molecular orbitals, etc. Our major concern has been the utilization of "affinity" between silicon and other hetero atoms to realize synthetically useful reactions. Of particular interest are the great bond dissocitation energies for Si—O [531 kJ/mol in $(CH_3)_3SiOCH_3$] and Si—F bonds [807 kJ/mol in $(CH_3)_3SiF$]. Utility of the eminent oxophilicity of silicon element has been amply demonstrated by various stoichiometric and catalytic reactions using trialkylsilyl triflates [1].

The affinity between silicon and fluorine is even stronger than the silicon to oxygen affinity. It is anticipated that, in this context, two ways are possible for activation of organic molecules, viz., combination of silicon-containing organic substrates and fluorine-based reagents and reaction between organofluorine compounds and silicon-based promoters. Firstly, fluoride anion can activate anionically certain organosilicon compounds by forming transient quinquecovalent silicon species (eq 1). In this connec-

$$R-SiR'_3 \; + \; F^- \; \rightleftharpoons \; R-SiR'_3F \; \rightleftharpoons \; R^- \; + \; R'_3SiF \quad (1)$$

tion, we have been intrigued by the chemistry of super-anionic phenoxide [2] and enolates [3]. Reaction of phenyl trimethylsilyl (TMS) ether and tris(diethylamino)sulfonium (TAS) difluorotrimethylsiliconate (a fluoride ion source) in THF produced TAS phenoxide. In a similar manner, the fluoride ion-promoted removal of TMS group from an enol silyl ether afforded the corresponding TAS enolate. The conductivity measurement and ^{1}H and ^{13}C NMR investigation have revealed that the anionic moieties are sufficiently naked in THF. Such species exhibited unique selectivities in their nucleophilic reactions.

$$TAS^+\ TMSF_2^-\ \rightleftharpoons\ TAS^+\ F^-\ +\ TMSF$$

$$TAS^+\ F^-\ +\ TMSOR\ \rightleftharpoons\ TAS^+\ RO^-\ +\ TMSF$$

$$RO = C_6H_5O\ \text{or}\ (\underline{Z})\text{-}C_6H_5CH=C(CH_3)O$$

RESULTS AND DISCUSSION

Certain classes of organofluorine compounds should have interact significantly with electron-accepting silicon species. thereby providing a new means for cationic activation of organic molecules according to eq 2.

$$R\text{--}F\ +\ SiX_4\ \rightleftharpoons\ R\text{--}\overset{+}{F}\text{--}Si\overset{-}{X}_4\ \rightleftharpoons\ R^+\ +\ FSiX_4^- \qquad (2)$$

To test the validity of this concept, we first studied a reaction of trityl fluoride and tetrafluorosilane. When a 1:1 mixture of trityl fluoride and a TMS carboxylate was exposed to a catalytic amount of tetrafluorosilane in acetonitrile, the corresponding trityl carboxylate was obtained in high yield [4]. TMS triflate was also usable as the catalyst. Similarly, the catalyzed reaction of trityl fluoride and phenyl TMS ether gave phenyl trityl ether.

Behavior of alcohols or their TMS ethers is highly dependent on their structures. 1-Octanol or octyl TMS ether underwent ordinary tritylation reaction in high yield. t-Butyl TMS ether was inert. Notably, cyclohexyl TMS ether was oxidized under the

$$(C_6H_5)_3CF\ +\ ROTMS\ \xrightarrow{\ \ \underset{\text{TMSOTf}}{\overset{SiF_4\ \text{or}}{}}\ \ }\ ROC(C_6H_5)_3\ +\ TMSF$$

R = acyl, phenyl, $\underline{pri}$-alkyl, etc.

standard tritylation conditions to give cyclohexanone in >80% yield
[5]. Interestingly, TMS ether of cis-4-t-butylcyclohexanol was
much more reactive than the trans isomer. The oxidation could
proceed via the corresponding trityl ethers. The tetrafluorosilane
catalyzed reaction of 3-cyclohexenyl TMS ether in acetonitrile
afforded a 2:1 mixture of 2-cyclohexenone and cyclohexanone.
All these phenomena are interpretable in terms of intermediary
trityl cation (yellow solution).

Glycosyl fluorides are now conveniently available among others
by treatment of 1-unprotected or 1-O-acetylated sugars with a
hydrogen fluoride—pyridine mixture [6]. We found that tetrafluorosilane or TMS triflate (<1 equiv) catalyzes effectively the
glycosylation using appropriately protected glycosyl fluorides and
TMS ethers (1:1 molar ratio) [7]. Table 1 exemplifies the new glycosylation procedure. Various fluorosugars and TMS ethers can be
used. Tetraalkoxysilanes or even unprotected alcohols as the nucleophiles are also usable in the tetrafluorosilane catalyzed reaction.

The present procedure using   silicon-based promotors is
advantageous over the classical Königs—Knorr glycosylation
method or the modified procedures which utilize precious and sometimes explosive silver salts or toxic mercury compounds. This
method is operationally simple and is suitable for the large-scale
reaction. Various functional groups are tolerable under the reaction conditions. Disaccharides have also been synthesized easily.
For example, the tetrafluorosilane catalyzed condensation of 2,3,4-
tri-O-acetyl-α-D-ribopyranosyl fluoride and 2,3,4-O-triacetyl-1-
O-p-styryl-β-D-glucopyranose in acetonitrile allowed a facile
entry to ptelatoside-A (9), a biologically interesting compound
isolated from bracken [8].

Particularly noteworthy is the characteristic solvent effect on
the steric course of the glycosylation with the glucopyranosyl
fluorides, 1α and 1β, which bear "nonparticipating" benzyloxy
group at the 2-position. Regardless of the stereochemistry of the
starting fluorides, the condensation with alkyl TMS ethers in
acetonitrile produced the β glucoside 2β with moderate to high
stereoselectivity, whereas the glycosylation in ether gave the α
anomer 2α predominantly. Since the products do not undergo
anomerization under   reaction conditions, the observed stereoselection is a result of kinetic control. In the absence of any
nucleophiles, the β fluoride 1β isomerized to the more stable α fluoride 1α rapidly in acetonitrile and rather slowly in ether. The
tetrafluorosilane catalyzed reaction is considered to occur via the
mechanism outlined in Scheme I. The stereoselectivity is controlled
by the equilibrium concentration and relative reactivity of the
intermediary ion-pairs. The α ion-pair 10α is thermodynamically
the more favored. However, anomeric 10β is more reactive than
10α, because the transition state is stabilized by a kinetic anomeric
effect. In polar acetonitrile (D 37.5), the ion-pairs are highly reactive and the stereoselectivity is determined mainly by the equilibrium concentration of the ion-pair intermediates, which results

Table 1 — Glycosylation via glycopyranosyl fluorides[a]

Nucleophile	Substrate	Catalyst (mol %)	Solvent[b]	% yield (α:β)
CH_3OTMS	1α	SiF_4 (50)	A	88 (15:85)
CH_3OTMS	1α	TMSOTf (100)	E	94 (84:16)
CH_3OTMS	1β	SiF_4 (50)	A	90 (16:84)
CH_3OTMS	1β	TMSOTf (90)	E	86 (90:10)
CH_3OTMS	3α	SiF_4 (40)	A	89 (58:42)
CH_3OTMS	3α	SiF_4 (c)	E	70[d] (66:34)
CH_3OTMS	4α	SiF_4 (30)	A	88 (22:78)
CH_3OTMS	4α	SiF_4 (c)	E	68[d] (33:67)
$(CH_3O)_4Si$[e]	1β	SiF_4 (5)	A	88 (16:84)
$(CH_3O)_4Si$[e]	1β	SiF_4 (c)	E	70[d] (74:26)
$c\text{-}C_6H_{11}OTMS$	1α	SiF_4 (30)	A	90 (15:85)
$c\text{-}C_6H_{11}OTMS$	1β	SiF_4 (20)	A	88 (15:85)
$c\text{-}C_6H_{11}OTMS$	1β	TMSOTf (65)	E	81 (86:14)
$c\text{-}C_6H_{11}OTMS$	1β	SiF_4 (30)	A + E	89 (15:85)
$c\text{-}C_6H_{11}OH$	1β	SiF_4 (20)	A	87 (16:84)
$t\text{-}C_4H_9OTMS$	1α	SiF_4 (40)	A	71 (23:77)
$t\text{-}C_4H_9OTMS$	1β	TMSOTf (40)	E	88 (79:21)
5	1α	SiF_4 (30)	A + E	85 (19:81)
5	1β	SiF_4 (c)	E	72[d] (77:23)
6	1α	SiF_4 (20)	A	73 (34:66)
6	1β	SiF_4 (c)	E	65[d] (67:33)
7	1α	SiF_4 (30)	A	89 (9:91)
7	1β	SiF_4 (20)	A	90 (9:91)
7	1β	SiF_4 (c)	E	68[d] (78:22)
8	1α	SiF_4 (30)	A	84 (22:78)
8	1β	SiF_4 (c)	E	66[d] (75:25)

[a] Reaction was carried out using equimolar amounts of a nucleophile and a substrate (0–25 °C, 0.5–24 h). [b] A = acetonitrile. E = diethyl ether. [c] Gaseous tetrafluorosilane was introduced into the mixture at 0 °C for 2 min. [d] Starting material was recovered in 20–30%. [e] Only 0.25 molar amount was used.

1α, X = F ; Y = H
1β, X = H ; Y = F

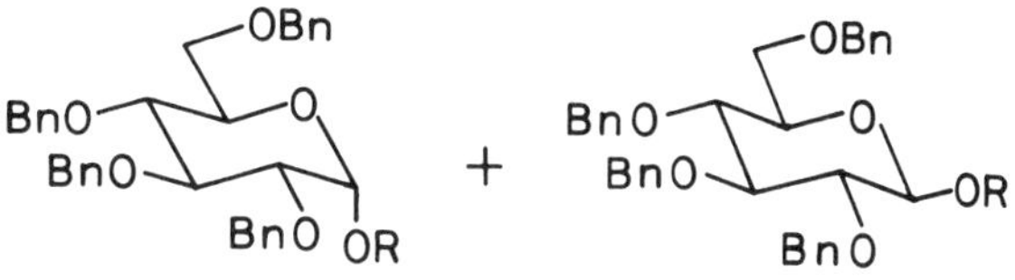

2α

2β

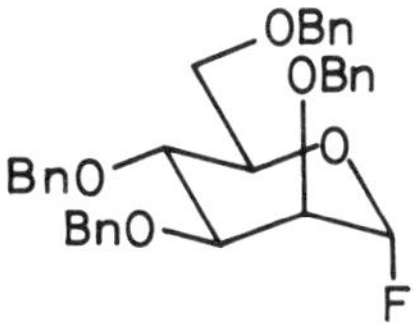

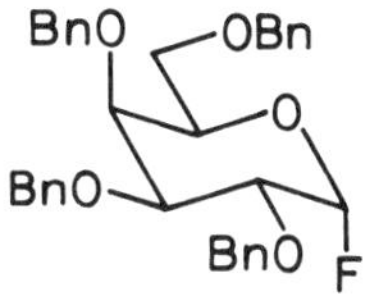

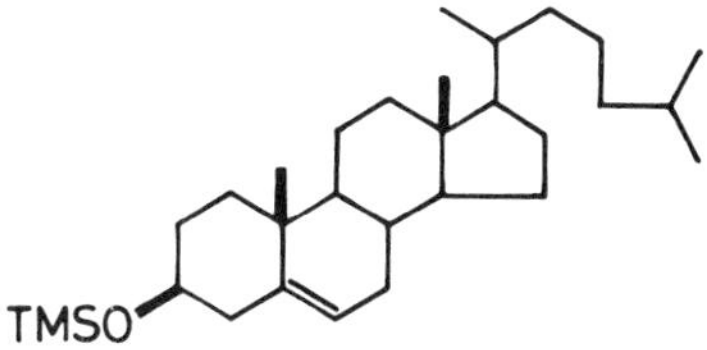

3α

4α

5

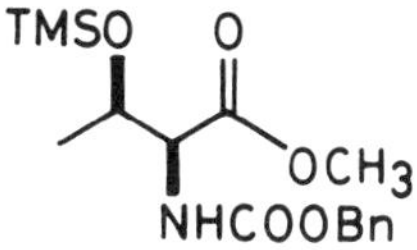

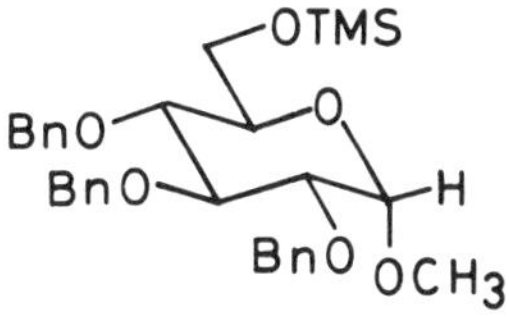

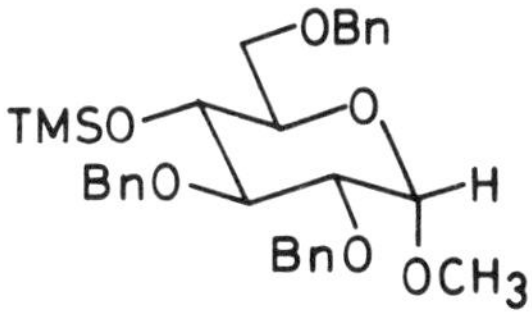

6

7

8

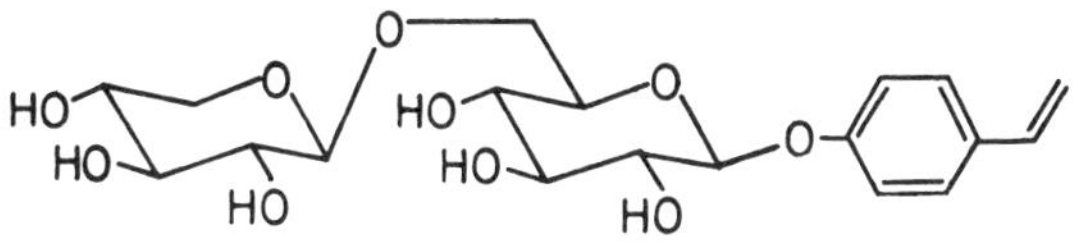

9

in the β-stereoselection. On the other hand, in ether ($\underline{D}$ 4.22) where the reaction is rather sluggish, the stereochemistry relies on the relative reactivities of the ion-pairs, favoring the α products. Donicity of the solvents seems unimportant in determining the stereoselectivity; the glycosylation in a 1:1 mixture of acetonitrile and ether exhibits a high degree of β-selectivity.

Scheme I

$$\alpha\text{-GlcF} \rightleftharpoons \alpha\text{-Glc}^+\text{SiF}_5^- \longrightarrow \beta\text{-GlcOR}$$
$$\underset{\underset{\sim}{1\alpha}}{} \qquad\qquad \underset{\underset{\sim}{10\alpha}}{} \qquad\qquad \underset{\underset{\sim}{2\beta}}{}$$

$$\updownarrow$$

$$\beta\text{-GlcF} \rightleftharpoons \beta\text{-Glc}^+\text{SiF}_5^- \longrightarrow \alpha\text{-GlcOR}$$
$$\underset{\underset{\sim}{1\beta}}{} \qquad\qquad \underset{\underset{\sim}{10\beta}}{} \qquad\qquad \underset{\underset{\sim}{2\alpha}}{}$$

REFERENCES

1. Review: Noyori, R.; Murata, S.; Suzuki, M. Trimethylsilyl Triflate in Organic Synthesis. Tetrahedron 1981, 37, 3899-3910.
2. Noyori, R.; Nishida, I.; Sakata, J. A Tris(dialkylamino)-sulfonium Phenoxide. Tetrahedron Lett. 1981, 22, 3993-3996.
3. Noyori, R.; Nishida, I.; Sakata, J. Tris(dialkylamino)sulfonium Enolates. Synthesis, Structure, and Reactions. J. Am. Chem. Soc. 1983, 105, 1598-1608.
4. Hashimoto, S.; Hayashi, M.; Noyori, R. An Easy Preparation of Triphenylmethyl Carboxylates. Bull. Chem. Soc. Jpn. 1984, 57, 1431-1432.
5. For related oxidation methods, see: Jung, M. E.; Speltz, L. Oxidation of Ethers via Hydride Abstraction: A New Procedure for Selective Oxidation of Primary, Secondary Diols at the Secondary Position. J. Am. Chem. Soc. 1976, 98, 7882-7884.
6. Hayashi, M.; Hashimoto, S.; Noyori, R. to be published.
7. Hashimoto, S.; Hayashi, M.; Noyori, R. Glycosylation Using Glucopyranosyl Fluorides and Silicon-Based Catalysts. Solvent Dependency of the Stereoselection. Tetrahedron Lett. 1984, 25, 1379-1382.
8. Ojika, M.; Wakamatsu, K.; Niwa, H.; Yamada, K.; Hirono, I. Isolation and Structures of Two New p-Hydroxystyrene Glycosides, Ptelatoside-A and Ptelatoside-B from Bracken, Pteridium Aquilinum var. Latiusculum, and Synthesis of Ptelatoside-A. Chem. Lett. 1984, 397-400.

CHAPTER 20

GROUP TRANSFER POLYMERIZATION. MECHANISTIC STUDIES

Dotsevi Y. Sogah and William B. Farnham Central Research and Development Department† E. I. du Pont de Nemours & Company, Inc. Experimental Station Wilmington, Delaware 19898 U.S.A.

INTRODUCTION

During the past three decades, there has been an increasing emphasis on development of synthetic methods for control of polymer architecture. "Living" anionic [1] and cationic [2] polymerization provide methodology for controlled polymerization of vinyl monomers. Despite the fact that anionic polymerization of butadiene and styrene has been carried out commercially for some time, commercial production of methacrylate polymers via the anionic method is hampered by requisite low temperatures needed to maintain "living" conditions. The process reported by Haggard and Lewis [1d] is operable at higher temperatures but it is only suitable for generation of methacrylate oligomers having average molecular weights of 500-3000.

Scheme 1

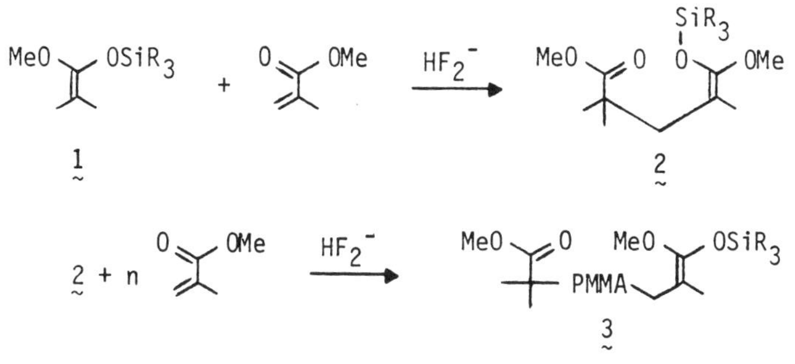

† Contribution No.

A fundamentally new method [3] that offers several advantages over existing processes was recently reported for polymerization of acrylic monomers. This method, termed group transfer polymerization (GTP), is a catalyzed Michael addition of silyl ketene acetals (1) [4] to α,β-unsaturated esters (Scheme I). The process is catalyzed by anions [3a,b] such as CN^-, N_3^-, HF_2^-, $Me_3SiF_2^-$, and selected Lewis acids [3c]. Methacrylate polymers with molecular weights as high as 50,000-100,000 are obtained at ambient temperature by GTP. In our preliminary communication [3a], we proposed an associative mechanism for this new process. This paper presents further evidence for such a mechanism.

Table 1 — Polymers by group transfer polymerization.

EXAMPLE NO.	MONOMER(S)	INITIATOR	CATALYST	SOLVENT	$\bar{M}_n$	$\bar{M}_w$	$\bar{M}_w/\bar{M}_n$	THEOR. M_n
1	Methyl Methacrylate	X	$TASHF_2$*	THF	10,200	11,900	1.17	10,100
2	Methyl Methacrylate	X	KHF_2	MeCN	18,000	21,400	1.18	20,200
3	Methyl Methacrylate	$\underline{1}^+$	$ZnBr_2$	$ClCH_2CH_2Cl$	6,020	7,240	1.20	3,400
4	Ethyl Acrylate	$\underline{1}^+$	$ZnBr_2$	$ClCH_2CH_2Cl$	17,000	26,600	1.57	10,100
5	Ethyl Acrylate	X	ZnI_2	CH_2Cl_2	3,300	3,400	1.03	3,360
6^a	Ethyl Acrylate	$\underline{1}^+$	$TASHF_2$	THF	27,200	59,400	2.16	26,100
7^b	Methyl Methacrylate (50%), n-butyl Methacrylate (50%)	X	$TASHF_2$	THF	8,010	8,550	1.07	8,200
8^c	Methyl Methacrylate (75%), 2-Ethylhexyl Methacrylate (25%)	$\underline{1}^+$	$TASHF_2$	THF	41,500	54,200	1.30	46,300
9^c	Lauryl Methacrylate (10%), Methyl Methacrylate (90%)	$\underline{1}^+$	$TASHF_2$	THF	6,540	7,470	1.14	7,000
10^c	Methyl Methacrylate (67%), 2-Ethylhexyl Methacrylate (33%)	$\underline{1}^+$	$TASHF_2$	THF	76,900	136,000	1.77	139,400

a) Run carried out at 0°C. b) Random copolymer, composition in mole percent. c) Block copolymer, composition in mole percent. †In 1, R = Me. * TAS = $(Me_2N)_3S^+$

$$X = \quad \begin{array}{c} Me_3SiO \\ Me_3SiO \diagdown \diagup_O \end{array} \Bigg\rangle = \Bigg\langle$$

RESULTS AND DISCUSSION

Characteristics of GTP

Group transfer polymerization of methyl methacrylate (MMA) is rapid, exothermic, and operable over a broad temperature range (-100° to +110°C), requiring only catalytic amounts (~0.01 to 0.1 mole percent relative to initiator) of the anionic catalysts. The polymers that are formed in quantitative yields have narrow molecular weight distribution, and the degree of polymerization is controlled by the ratio of monomer to initiator. The polymers are living and, consequently, further addition of the same monomer leads to the expected increase in molecular weight; sequential addition of different monomers gives block copolymers (Table 1, entries 8-10). GTP is compatible with a variety of functional groups [3a,b,c], and incorporation of a terminal functionality is facile [3d].

Mechanistic studies

There are several mechanisms that might be written for GTP. The principal mechanistic question that we choose to address is whether GTP takes place by the dissociative mechanisms shown in Schemes II and III or by the associative mechanism of Scheme IV.

Scheme II — Reversible dissociative mechanism

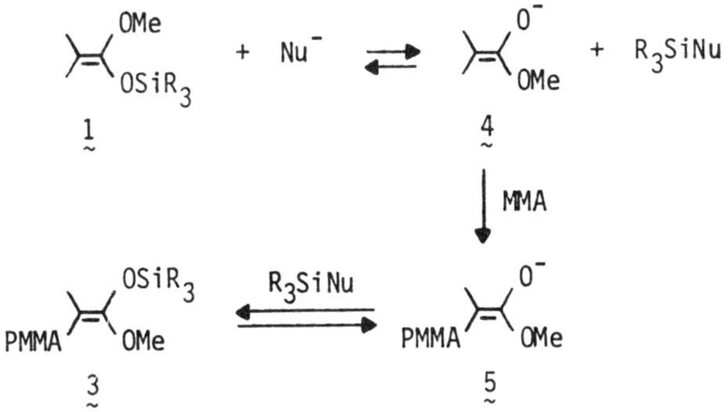

Scheme III — Irreversible dissociative mechanism

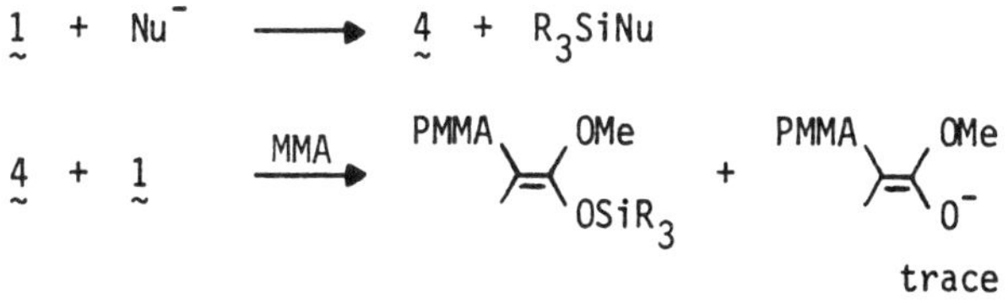

Scheme IV − Associative mechanism

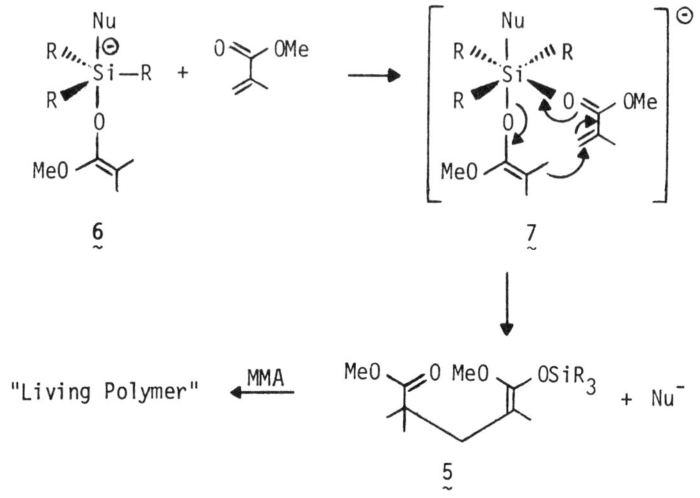

In the first dissociative mechanism, the trialkylsilyl fragment is cleaved in a reversible step to give R_3SiNu and an ester enolate (4). The latter then undergoes repeated Michael addition to MMA to provide a polymeric ester enolate (5) which, upon resilylation by R_3SiNu, gives the observed polymeric silyl ketene acetal (3).

Scheme III shows a second dissociative mechanism which is similar to II except that cleavage of R_3Si is irreversible and the resilylation steps are carried out by other silyl ketene acetals (e.g. 1) present in the reaction [5]. Since the role of liberated R_3SiNu is quite different in the two dissociative processes, experiments involving possible incorporation of labelled R_3SiNu species (vide infra) can exclude II; but double-labelling experiments (vide infra) are required to exclude III.

In the associative mechanism (Scheme IV), the nucleophilic catalyst coordinates to the silicon atom to provide a pentacoordinate species (6) [6a]. This intermediate then reacts with MMA forming new C-C and Si-O bonds, and cleaving the old Si-O bond. The important distinction in this mechanism is that silyl exchange among growing chains is excluded; hence, the identity of the silicon atom of the initiator molecule remains invariant throughout the growth of the polymer chain. We have proposed the involvement of a hexacoordinate silicon species (7) [3a,6b], recognizing that the viability of such a hypothesis depends critically upon the timing of the above-mentioned bond-forming and bond-breaking events. We have sought to test the "hexacoordinate intermediate" hypothesis with two different approaches (vide infra) which were hoped to be sufficiently "risky". Our

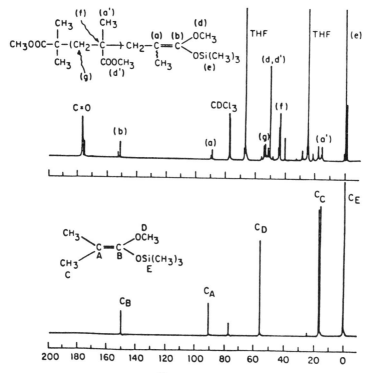

Figure 1 — 100.6 MHz ^{13}C spectra of "living" PMMA and 1.

experimental approach involves isolation and identification of the living polymer, and examination of labelled exchange reactions, solvent and temperature dependence, and polymer tacticity.

Isolation and characterization of intermediates

Figure 1 shows the 100.6 MHz ^{13}C-nmr spectra of the isolated living oligomer and silyl ketene acetal initiator. The distinct sp^2 carbon (C_A, C_B) and Me$_3$Si (C_E) signals of the initiator match the corresponding ones [C(a), C(b), C(e)] of the oligomer (d.p. ~10). In a separate experiment, treatment of the GTP reaction mixture with benzaldehyde gave β-hydroxy ester end group (9, Scheme V), while phenyl isocyanate gave an amide end group (10, Scheme V). These results, together with our earlier report on benzylation of group transfer polymers [3d], clearly demonstrate that the polymers, indeed, contain silyl ketene acetal functional groups.

Scheme V

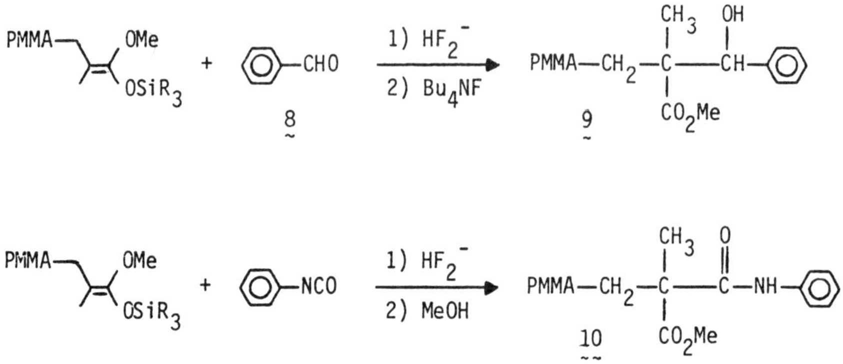

$$PMMA{-}\underset{OSiR_3}{\overset{OMe}{\diagdown}} + \bigcirc{-}CHO \quad \xrightarrow[\text{2) } Bu_4NF]{\text{1) } HF_2^-} \quad PMMA{-}CH_2{-}\underset{CO_2Me}{\overset{CH_3}{\underset{|}{\overset{|}{C}}}}{-}\underset{}{\overset{OH}{\underset{|}{CH}}}{-}\bigcirc$$

8   9

$$PMMA{-}\underset{OSiR_3}{\overset{OMe}{\diagdown}} + \bigcirc{-}NCO \quad \xrightarrow[\text{2) } MeOH]{\text{1) } HF_2^-} \quad PMMA{-}CH_2{-}\underset{CO_2Me}{\overset{CH_3}{\underset{|}{\overset{|}{C}}}}{-}\underset{}{\overset{O}{\overset{||}{C}}}{-}NH{-}\bigcirc$$

10

The question arises as to whether or not GTP is reversible since Michael reactions, under certain conditions, are reversible [7]. To address this question, two living polymers, one having $\bar{M}_n$ 1590, $\bar{M}_w$ 1680, D 1.06, and the other $\bar{M}_n$ 3750, $\bar{M}_w$ 3870, D 1.01, were mixed and stirred overnight in the presence of tris(dimethylamino)sulfonium (TAS) bifluoride. The gel permeation chromatogram of the pure, isolated polymers was bimodal with each maximum corresponding to the molecular weight of each of the individual initial polymers, implying that the living polymers did not undergo reversible depolymerization. This result, coupled with the fact that quantitative yields of monodisperse polymers are obtained, suggests that GTP is not reversible. Having established the above facts, we then turned our attention to studies that exclude the two dissociative mechanisms (Schemes II and III).

Silyl fluoride labelling studies

To provide evidence for the absence (or presence) of intermediate fluorosilanes in the GTP reaction, labelled silyl fluoride exchange experiments were performed. To successfully do this, catalyst quenchers capable of trapping the catalysts, thereby stopping the reaction, are needed. Tetracoordinate silane, 11, was used as a catalyst quencher for $Me_3SiF_2^-$ catalyzed GTP since 11 irreversibly binds fluoride ion, forming an inert pentacoordinate silicate [8]. The corresponding TAS bifluoride catalyzed GTP is quenched by silver nitrate since $TASNO_3$ is not a GTP catalyst in THF.

$$TASHF_2 + AgNO_3 \longrightarrow AgHF_2 + TASNO_3$$

Table 2 summarizes the results of our studies. Polymerization of MMA with phenyldimethylsilyl initiator (12A) in the presence of an equimolar quantity of tolyldimethylsilyl fluoride (Ar' = Tol) with $TASHF_2$ catalyst (25-75°C,

Table 2. Silyl Fluoride Labelling Studies

$$\underset{1.0}{\overset{OSiMe_2Ar}{\diagdown}\!\!\!=\!\!\!\underset{OMe}{\diagup}} \;+\; \underset{1.0}{Ar'SiMe_2F} \;+\; \underset{15}{MMA} \;\xrightarrow{\text{cat}}\; \underset{PMMA}{\overset{OSiMe_2Ar}{\diagdown}\!\!\!=\!\!\!\underset{OMe}{\diagup}}$$

12A Ar = Ph   13A Ar = Ph
  B Ar = Tol   B Ar = Tol

Catalyst	Temp/Time	Ar	Ar'	Polymer End-group Analysis[†]
TASHF$_2$	25-55°	Ph	Tol	No detectable TolSiMe$_2$
TASHF$_2$	25-55°	Tol	Ph	<10% PhSiMe$_2$
TASMe$_3$SiF$_2$	-70°/0.5 h*	Ph	Tol	50/50 PhSiMe$_2$/TolSiMe$_2$
TASMe$_3$SiF$_2$	-70°/5 min*	Ph	Tol	67/33 PhSiMe$_2$/TolSiMe$_2$
TASMe$_3$SiF$_2$	-90°/5 min*	Ph	Tol	ca. 90/10 PhSiMe$_2$/TolSiMe$_2$

[†] ^{1}H nmr (360 MHz).
* Total contact time of silyl ketene acetal, silyl fluoride, MMA
and catalyst before quench with spiro silane 11.

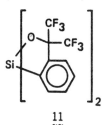

11

0.75 hr.) provided an oligomer (d.p. ~15) which contained no
detectable tolyldimethylsilyl end group. Similarly, use of
12B and phenyldimethylsilyl fluoride (Ar' = Ph) gave less
than 10% of the exchanged product. Similar results were
obtained with corresponding trimethylsilyl and triethylsilyl
compounds. With Me$_3$SiF$_2^-$ catalyst, however, very low temperatures and short reaction times were required to suppress
the rate of silyl group exchange. These results show that
under GTP reaction conditions, fluorosilanes are not produced
in a reversible, dissociative step; hence, the reversible
dissociative mechanism (Scheme II) is excluded.
     Since GTP requires only minute amounts of the nucleo-
philic catalyst (<<1%), the concentration of the initiator
far exceeds that of the hypothesized dissociated species
(Scheme III). Hence, resilylation of the ester enolate by
other silyl ketene acetals (monomeric or oligomeric), may be
facile and pronounced at the early stages of the reaction.
This is corroborated by our finding that living oligomers and
silyl ketene acetals readily exchange silyl groups in the

presence of GTP catalysts (Scheme VI); on the other hand, the living oligomers do not readily exchange silyl groups with one another under identical conditions (Scheme VII).

Scheme VI

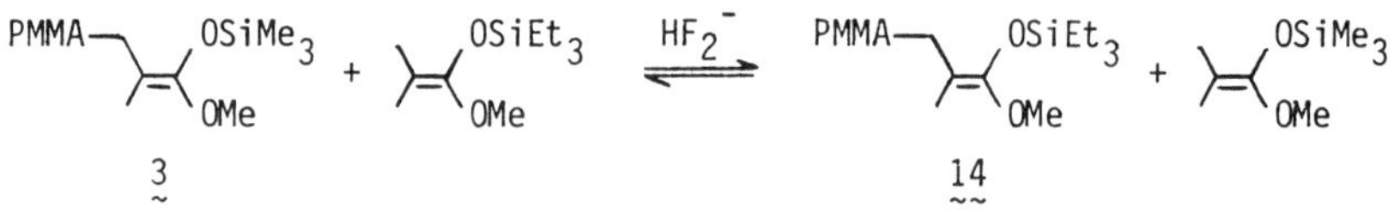

Scheme VII

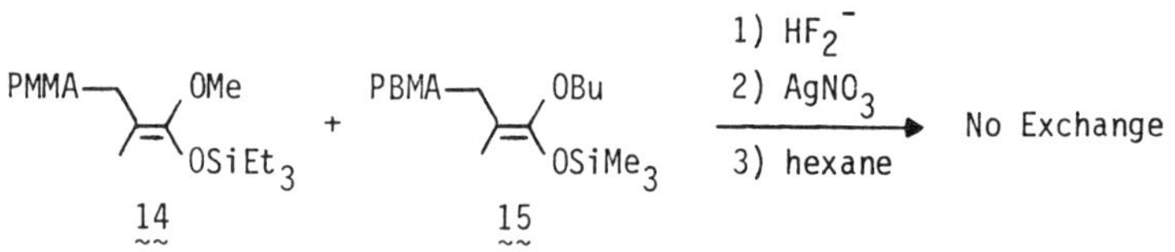

Therefore, the important mechanistic question lies in the difference between the relative rates of silyl group exchange and intramolecular silyl transfer during propagation. To shed more light on this question, appropriate double-labelling experiments were performed.

Double-labelling studies

A mixture of living PMMA (14) and poly(n-butyl metha-crylate) (PBMA) (15) was used to initiate GTP of n-butyl methacrylate (BMA) in the presence of bifluoride ion (Scheme VIII). The reaction was quenched with $AgNO_3$ when about 70% of the added BMA was consumed. The precipitated $AgHF_2$ was removed by filtration after centrifugation.

Scheme VIII

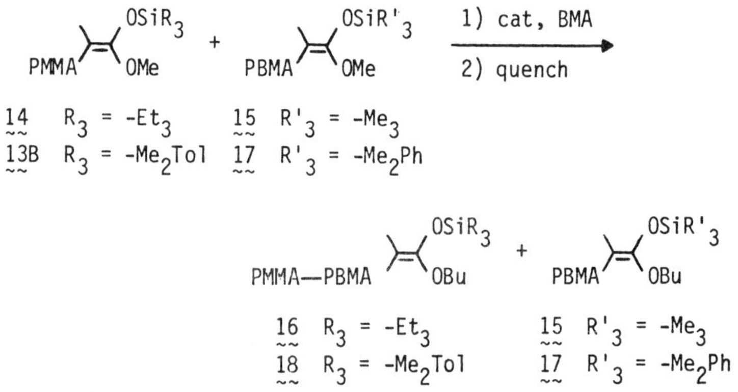

Treatment of the filtrate with dry hexane under nitrogen atmosphere precipitated the PMMA-PBMA block copolymer which contained only the triethylsilyl ketene acetal end group (16). The hexane-soluble PBMA contained only the trimethylsilyl ketene acetal functional group. Both polymers had molecular weights expected from the relative amounts of oligomers used and BMA consumed. No cross-over products were detected. Similar experiments using living oligomers labelled with dimethyltolylsilyl (13B) and dimethylphenylsilyl (17) groups gave living group transfer polymers with no detectable silyl exchange. The corresponding experiments with Me$_3$SiF$_2^-$ as catalyst at -90°C also showed that polymerization occurred without silyl exchange after quenching with spirosilane 11. We conclude that the propagation rate exceeds that of silyl exchange among living oligomers.

The results of our double-labelling experiments support an "intramolecular" silyl transfer mechanism in which the silyl group is transferred from the oxygen of the initiator to the carbonyl oxygen of the incoming monomer, possibly via hypervalent silicon intermediates, 6 and 7. Since the rate of silyl exchange between initiator and living oligomer (vide supra) and propagation rate are competitive, our labelling experiments alone cannot rule out intervention of the irreversible dissociative mechanism (Scheme III) during the early stages of the reaction. We believe that the mechanism for the propagation of GTP is independent of the degree of polymerization and that a change in mechanism as the reaction progresses is incompatible with the fact that polymer molecular weight is controlled by the initiator.

Polymer tacticity: solvent, temperature and catalyst dependence

Anionic polymerization of MMA initiated by alkyllithium reagents gives PMMA whose tacticity is determined by the polarity of the polymerization media. Highly isotactic polymers are obtained in non-polar solvents (e.g., toluene), while polar solvents (e.g., THF) give syndiotactic polymers [9]. However, tacticity of PMMA made by GTP is independent of solvent but dependent upon the reaction temperature and nature of the catalyst. Polymers by anion-catalyzed GTP made at ambient temperature possess syndiotactic and heterotactic sequences in the ratio of 55:45, respectively, with no measurable isotactic component in all solvents examined. The syndio-selectivity increases as the temperature is lowered with syndiotactic:heterotactic ratios of 2:1-4:1. Syndiotactic:heterotactic ratios >2:1 are obtained in Lewis acid-catalyzed GTP regardless of solvent and temperature [3c]. Lewis acids are believed to operate by activating monomer through coordination to its carbonyl functional group. The observed tacticity of polymers made by GTP as a function of

temperature, solvent and catalyst, vis-a-vis results obtained
by anioinic polymerization, militates against the involvement
of dissociative processes.

Model studies

In our proposed associative silyl transfer mechanism,
the nucleophilic catalyst activates the intiator by coordination to silicon. To test the involvement of the hypothesized
hypervalent species by means of a model system, we made
recourse to a special ligand known to provide stability to
pentacoordinate silicon structures [8a,10]. We hypothesize
that a pentacoordinate silyl enolate intermediate should
react with MMA without added nucleophilic catalysts.
Treatment of silane 11 with the lithium enolate of
methyl isobutyrate provided a stable pentacoordinate silicate, 19 (Scheme IX) [11]. Silicate 19, indeed, reacted with
methyl methacrylate (at ca. 25-65°C) without added catalyst
to give PMMA of reasonably narrow molecular weight distribution. While the ligands of 7 and 19 are admittedly
different, we regard this result as corroborative evidence
for the involvement of pentacoordinate silicon species in
GTP. The results also suggest that coordinative activation
of the initiator is a necessary although not necessarily a
sufficient step in the process.

Scheme IX

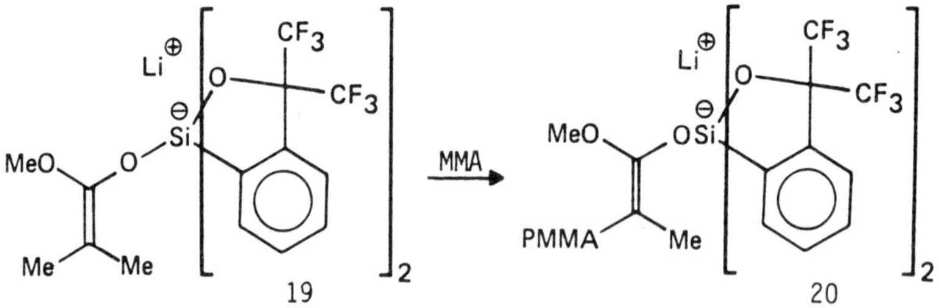

19   20

The intermediacy of hexacoordinate structure 7 would
require retention of configuration at silicon during the
transfer step. We synthesized both diastereomers of sila-
cyclopentane intiators (21A and B). Unfortunately, these
cyclic silyl ketene acetals undergo stereomutation at silicon
under GTP reaction conditions. We are investigating other
cyclic systems.

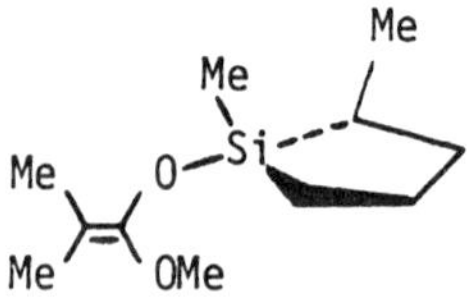

21A cis
 B trans

CONCLUSION

Group transfer polymerization has been shown to offer the following advantages over existing processes: operability over a broad temperature range, wide choice of solvents, excellent molecular weight control, compatibility with a variety of functional groups, facile incorporation of terminal functional groups, and block and random copolymer synthesis. We have provided evidence that the process occurs via an associative rather than a dissociative mechanism. By showing that added fluorosilanes do not exchange with living polymers and that the oligomeric silyl ketene acetals do not exchange silyl groups intermolecularly, we have eliminated the possibility that dissociative mechanisms could be operating exclusively. The above facts, together with the results of our model studies, support our proposed associative mechanism.

ACKNOWLEDGEMENT

We thank B. E. Smart, B. M. Trost, T. Fukunaga, T. V. RajanBabu and O. W. Webster for useful discussions. We also thank T. Anemone for the manuscript preparation.

REFERENCES

1. (a) Morton, M. Anionic Polymerization: Principles and Practice. Academic Press, 1983. (b) Szwarc, M. Carbanions, Living Polymers and Electron Transfer Processes. Wiley-Interscience, New York, 1968. (c) Noshey, A.; McGrath, J. E. Block Copolymers. Academic, New York, 1977. (d) Haggard, R. A.; Lewis, S. N. Prog. Org. Coatings, 1984, $\underline{12}$, 1-26.
2. Mandal, B. M.; Kennedy, J. P. J. Polym. Sci., Polymer Chem. Ed., 1978, $\underline{16}$, 833.
3. (a) Webster, O. W.; Hertler, W. R.; Sogah, D. Y.; Farnham, W. B.; RajanBabu, T. V. J. Amer. Chem. Soc., 1983, $\underline{105}$, 5706 and references cited therein. (b) Webster, O. W. U.S. Patent 4,417,034; Farnham, W. B.; Sogah, D. Y. U.S. Patent 4,414,372. (c) Hertler, W. R.; Sogah, D. Y.; Webster, O. W. Macromolecules, 1984, (in press). (d) Sogah, D. Y.; Webster, O. W. J. Polym. Sci., Polym. Lett. Ed. 1983, $\underline{21}$, 927.
4. (a) Ainsworth, C.; Chen, F.; Kuo, Yu-Neng J. Organomet. Chem., 1972, $\underline{46}$, 59. (b) Colvin, E. Silicon in Organic Synthesis. Butterworths, London, 1981. (c) Brownbridge, P. Synthesis, 1983, 1; ibid, 1983, 85.
5. RajanBabu, T. V. J. Org. Chem., 1984, $\underline{49}$, 2083.
6. (a) Noyori, et al. have proposed, without experimental support, pentacoordinate silicon species as reactive

intermediates in the fluoride ion-catalyzed addition of silyl enol ethers to carbonyl compounds. Subsequent papers from this group tend to favor other mechanisms for these reactions. See J. Am. Chem. Soc., 1977, 99, 1265; ibid 1980, 102, 1223.
(b) Corriu has also proposed involvement of hexacoordinate silicon intermediates in several nucleophile-catalyzed organosilicon reactions. See Corriu, R. J. P.; Perez, R.; Reye, C. Tetrahedron, 1983, 39, 999 and references cited therein. For other hexacoordinate species, see Farnham, W. B.; Whitney, J. F. J. Amer. Chem. Soc., 1984 (in press); Kumer, Von D.; Koster, H. Z. Anorg. Allg. Chem., 1973, 398, 279; Kummer, Von D.; Balkir, A; Koster, H. J. Organomet. Chem., 1979, 178, 29.

7.   House, H. O. Modern Synthetic Reactions. W. A. Benjamin, Phillipines, 1972, 595-623.

8.   (a) Farnham, W. B.; Harlow, R. L. J. Amer. Chem. Soc., 1981, 103, 4608; Farnham, W. B.; Whitney, J. F. ibid, 1984 (in press); (b) Perkins, C. W.; Martin, J. C.; Arduengo, A. J.; Lau, W.; Alegria, A; Kochi, J. K. J. Am. Chem. Soc., 1980, 102, 7753. (c) Michalak, R. S.; Martin, J. C. ibid, 1982, 104, 1683. (d) Font Freide, J. J. H. M.; Trippett, S. J. Chem. Soc., Chem. Comm., 1980, 934. (e) Totsch, W.; Sladky, F. ibid, 1980, 927.

9.   (a) Fox, T. G.; Garret, B. S.; Goode, W. E.; Gratch, S.; Kincaid, J. F.; Spell, A.; Stroupe, J. D. J. Am. Chem. Soc., 1958, 80, 1768. (b) Glusker, D. L.; Evans, R. A. J. Am. Chem. Soc., 1964, 86, 187. (c) Yuki, H.; Hatada, K.; Kikuchi, Y.; Niinomi, T. J. Polym. Sci., Polym. Lett., 1968, 6, 753. (d) Allen, P. E. M.; Mair, C.; Fisher, M. C.; Williams, E. H. J. Macromol. Sci. - Chem. 1982, A17, 61.

10.   Martin, J. C.; Perozzi, E. F. J. Am. Chem. Soc., 1974, 96, 3155; Perozzi, E. F.; Martin, J. C. ibid, 1979, 101, 1591.

11.   Silicate 19 is the first example of a bona fide pentacoordinate silyl enolate.

CHAPTER 21

SILAFUNCTIONAL COMPOUNDS IN ORGANIC SYNTHESIS

Kohei Tamao Department of Synthetic Chemistry, Kyoto University, Yoshida, Kyoto 606, Japan

INTRODUCTION

Electrophilic cleavage of the carbon–silicon bond is the key step in the synthetic applications of organosilicon compounds. The silicon reagents so far used have been restricted to those which contain "activated" carbon–silicon bonds, such as allyl–, vinyl–, aryl–, ethynyl–, and allenyl-trimethylsilanes. Cleavage reactions of "unactivated" carbon–silicon bonds, such as alkyl–silicon bonds, have scarcely been used in synthetic organic chemistry [1].

Our recent effort has been devoted to finding some new methodologies for the electrophilic cleavage of the unactivated carbon–silicon bond, particularly in the field of silafunctional organosilicon compounds. This review is concerned with their unique and important role in functional group transformations.

WHY SILAFUNCTIONAL COMPOUNDS?

There are several factors in electrophilic cleavage of the carbon–silicon bonds: ① reactivity of the relevant R group, ② electronic and steric effects of the remaining R' groups, ③ chemoselectivity between R and R' groups, ④ activation of the electrophiles, and ⑤ nucleophilic assistance on silicon.

In view of these factors, electrophilic cleavage reactions
will be analyzed briefly. It has been well accepted that nucleophilic substitution reactions at silicon proceed through
pentacoordinate or .sometimes hexacoordinate silicon intermediates [2]. This means that such extracoordinate silicon
species should be involved and play an important role also in
electrophilic cleavage of the carbon–silicon bonds. This nucleophilic assistance however appears to have little been taken
into account in most of the previous works, which have dealt
with cleavage of reactive carbon–silicon bonds by an activated
electrophile together with a weak nucleophile. In such cases,
the reaction is frequently described as if it were initiated by
the electrophilic attack on the reactive organic group in the
first step, followed by the nucleophilic attack on silicon in
the second step, as shown in eq. (1).

There should be a reverse sequence of reactions. Thus,
less reactive carbon–silicon bonds may be activated by extra-
coördination via nucleophilic attack on silicon in the first
step and cleaved by an electrophile in the second step, as shown
in eq. (2).

$$R'_3Si\text{-}R \left\{ \begin{array}{l} \xrightarrow{\ E^{(+)}\ } [R'_3Si\text{—}R\text{—}E]^{(+)} \xrightarrow{\ Nu^{(-)}\ } R'_3Si\text{-}Nu \ + \ R\text{-}E \qquad (1) \\ \\ \xrightarrow{\ Nu^{(-)}\ } [Nu\text{—}R'_3Si\text{—}R]^{(-)} \xrightarrow{\ E^{(+)}\ } R'_3Si\text{-}Nu \ + \ R\text{-}E \qquad (2) \end{array} \right.$$

A contrastive and complementary electronic effect may be
envisaged in the above two routes. Thus, in the former route,
the presence of electronegative groups on silicon should
diminish the reactivity of the carbon–silicon bond, while in the
latter case, the more the electronegative groups on silicon, the
easier the formation of hypervalent, extracoordinate silicon
species, and in turn the easier the cleavage reaction. The
latter route should be realized not with ordinary organotrimethylsilanes, but with silafunctional organosilicon compounds.

In addition to the activation of the carbon–silicon bond
mentioned above, extracoordination would bring about activation
of electrophiles through inclusion to the coordination sites
[3]. The former may well be designated as a hyper–valency
effect and the latter as a template effect. It is these two
effects that are expected of silafunctional compounds.

Silafunctional compounds have several advantages over
organotrimethylsilanes. (1) A variety of functional groups,
such as hydro, fluoro, chloro, amino, alkoxy groups and so on,
can be introduced onto silicon. (2) The number of the functional groups may be chosen appropriately. (3) Interchange

between and incorporation of different functional groups are possible. (4) Additional functions can be exhibited through structural design of functional groups, exemplified by the use of optically active amino or alkoxy groups.

Consequently, silafunctional organosilicon compounds promise many possibilities, which are impossible with traditional trimethylsilyl derivatives, and may open a new field in synthetic organic chemistry.

STEREOSELECTIVITY CONTROL IN HALOGEN CLEAVAGE OF ALKENYL-SILICON COMPOUNDS

The most typical valency effect has been observed in stereochemical studies on halogen cleavage of alkenyl-silicon compounds, the silyl groups being $SiMe_3$, SiF_3, and SiF_5^{2-} [4]. It has been shown that retention may result from halogen attack on extracoordinate silicon species, while inversion may arise from halogen addition to ordinary tetracoordinate alkenylsilanes followed by treatment with nucleophiles.

Based on these observations, a one-pot stereoselective synthesis of (E)- and (Z)-alkenyl bromide from acetylene has been developed via hydrosilylation with triethoxysilane (eqs. 3 and 4) [5]. Noteworthily, the complete stereochemical control is attained with a common (E)-alkenylsilane by merely changing the order of treatment with KHF_2 and halogen.

$$ R-C \equiv C-R \ + \ HSi(OEt)_3 \ \xrightarrow{H_2PtCl_6} \ \underset{R}{\overset{R}{\diagdown}} Si(OEt)_3 $$

$$ \xrightarrow[\text{MeOH}]{(1)\ KHF_2}\ \text{silicate} \ \xrightarrow[\text{or NBS}]{(2)\ Br_2}\ \underset{R}{\overset{R}{\diagdown}}Br \qquad (3) $$

$$ R = C_4H_9 \quad 70\text{-}80\% $$

$$ \xrightarrow[\text{CCl}_4]{(1)\ Br_2}\ \text{adduct} \ \xrightarrow[\text{MeOH}]{(2)\ KHF_2}\ \underset{R}{\overset{Br}{\diagdown}}R \qquad (4) $$

OXIDATIVE CLEAVAGE OF THE CARBON-SILICON BOND IN SILAFUNCTIONAL ORGANOSILICON COMPOUNDS

A priori, two methodologies may be envisaged for the introduction of an oxygen functionality into an organic group via cleavage of the carbon-silicon bond (eqs. 5 and 6) [6]. (Hereafter, si represents unspecified silyl moieties).

Thus, the oxidizing agent interacts firstly either with the organic group rather than the silyl group (route A) or with the

silicon center rather than the organic group (route B). All of
the procedures so far reported have involved the organic groups
more of less activated towards oxidizing agents and have been
based on the principle of route A. In route B, the silyl group
activated towards oxidizing agents acts as a template for the
oxidation. Route B therefore would be widely applicable to any
kind of oganic groups if only linked to an appropriate silafunctional group.

$$A \left(\begin{matrix} [O] \\ [O] \end{matrix} \right) B \qquad A \nearrow \qquad \overset{[O]}{\underset{|}{R-si}} \longrightarrow R\text{-}0\text{-}si \longrightarrow R\text{-}OH \qquad (5)$$

$$R-si \qquad \qquad B \searrow \qquad \overset{[O]}{\underset{|}{R-si}} \longrightarrow R\text{-}0\text{-}si \longrightarrow R\text{-}OH \qquad (6)$$

A typical and fundamental example of this route is an
intramolecular rearrangement of triorganosilyl perbenzoates to
alkoxysilanes, reported by Buncel and Davies in 1958 [7].
Little attention, however, has been paid to the synthetic application of this type of reaction until recently.
    We have found that the carbon-silicon bonds in sila-
functional compounds are readily cleaved by several types of
oxidizing agents to form the corresponding alcohols or carbonyl
compounds, respectively from alkyl- or alkenyl-silanes.

A. Oxidation of silafunctional alkyl- and aryl-silanes

    Table 1 summarizes typical raction conditions [A] - [G] and
silyl groups which are capable of being converted to a OH group.
Suitable reaction condition(s), if not necessarily been
optimized yet, are listed for each silyl group. Yields obtained
with n-octyl derivatives were in the range of 70-100%.
Evidently, we are in a position to be able to convert most of
the common silafunctional groups to the hydroxy group [6,8,9].
    Stereochemistry at aliphatic carbon is completely retained
regardless of the reaction conditions, as shown by the oxidation of exo- and endo-2-norbornyl silicon compounds (Scheme 1).
    Worthy of note is the "direct" introduction of the OH group
regioselectively onto the carbon atom to which the silicon has
been attached. This oxidation thus provides the first successful procedure for the conversion of allylsilanes to allyl
alcohols without an allylic transposition (eq. 7) [10].

$$R\diagdown\!\diagup\!\diagdown\text{SiMe}(0\text{-}i\text{-Pr})_2 \xrightarrow{\text{[E]}} R\diagdown\!\diagup\!\diagdown\text{OH} \quad (\text{no} \quad \overset{R\diagdown}{\underset{OH}{\diagup\!\diagdown}}) \quad (7)$$

Table 1 — Typical reaction conditions (A—G) for oxidation of silafunctional alkyl- and aryl-silanes and silyl groups amenable to conversion to OH group.

$$R\text{-}si \quad \rightleftharpoons \quad R\text{-}OH$$

[A]	MCPBA / DMF / r.t.
[B]	MCPBA / cat. KF / DMF / r.t.
[C]	MCPBA / xs. KF / DMF / r.t.-60°C
[D]	90% H_2O_2 / KF / DMF / r.t.-60°C
[E]	30% H_2O_2 / KHF_2 / DMF / 60°C [neutral]
[F]	30% H_2O_2 / Ac_2O / KHF_2 / DMF / r.t. [acidic]
[G]	30% H_2O_2 / $KHCO_3$ or $NaHCO_3$ / MeOH / THF / 60°C [basic]

$SiMe_2H$	[G]	$SiMe_2(NR'_2)$	[G]
$SiMe_2F$	[C] [G]	$SiMe_2(OR')$	[E] [G]
$SiMeF_2$	[B]	$SiMe(OR')_2$	[C] [D] [E] [F] [G]
SiF_3	[A]	$Si(OR')_3$	[C] [D] [F] [G]
SiF_5^{2-}	[A]	$SiMe_2OSiMe_2R$	[D] [G]^a
$SiMe_2Cl$	[G]	$SiMe_2(CH_2CH=CH_2)$	[G]^a
$SiMeCl_2$	[G]	$SiMe_2Ph$	[G]^a
$SiCl_3$	[G]		

^{a} The siloxane, allyl-silicon, and phenyl-silicon bonds should be cleaved with CF_3COOH/KHF_2 prior to the oxidation. This two-step procedure can be performed in a single flask.

Scheme 1

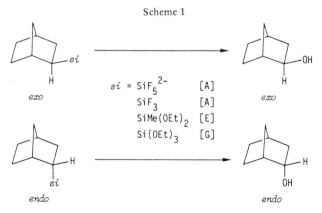

$$si = SiF_5^{2-} \quad [A]$$
$$SiF_3 \quad [A]$$
$$SiMe(OEt)_2 \quad [E]$$
$$Si(OEt)_3 \quad [G]$$

These oxidative cleavage reactions are considered to proceed through the intramolecular migration of an organic group from silicon to oxygen in extracoordinate silicon intermediates. A simplified possible mechanism for the hydrogen peroxide oxidation is represented in Scheme 2.

Scheme 2

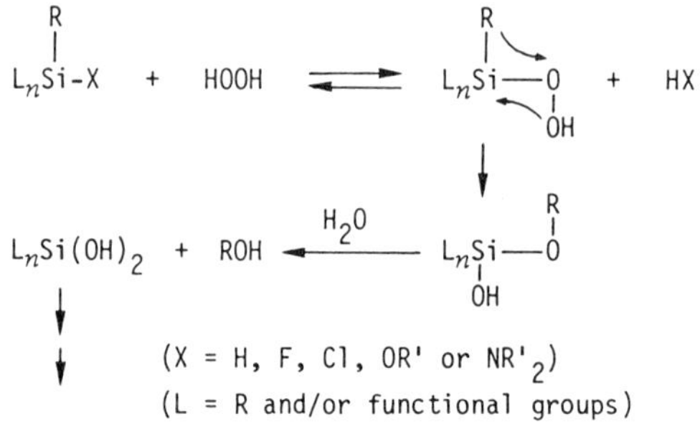

$$R_nSi-X \ + \ HOOH \ \rightleftarrows \ L_nSi-O \ + \ HX$$

$$L_nSi(OH)_2 \ + \ ROH \ \xleftarrow{\ H_2O\ } \ L_nSi-O$$

(X = H, F, Cl, OR' or NR'$_2$)

(L = R and/or functional groups)

B. Oxidation of silafunctional alkenylsilanes

Oxidation of silafunctional alkenylalkoxysilanes, like alkenyltrifluorosilanes [11], provides new routes to carbonyl compounds [12]. Since the present oxidation may proceed through an enol silyl ether species (eq. 8), several new types of transformations have been realized, which have been impossible by a well-known standard method via a sequence of epoxidation and acidolysis [13]. Of particular interest are carboxylic acid

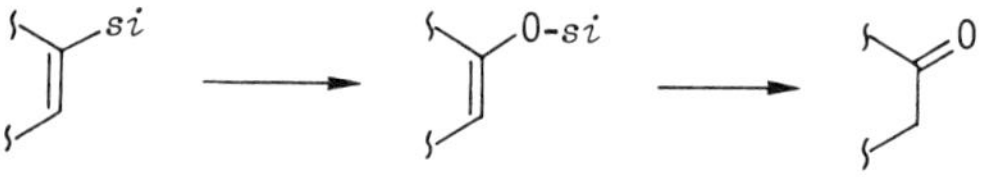

$$\text{(8)}$$

synthesis (eq. 10), synthesis of α-hydroxyketones via epoxidation and the subsequent oxidative cleavage (eq. 12) [5], and the direct transformation of cyclic vinylsilanes to ketones (eq. 13).

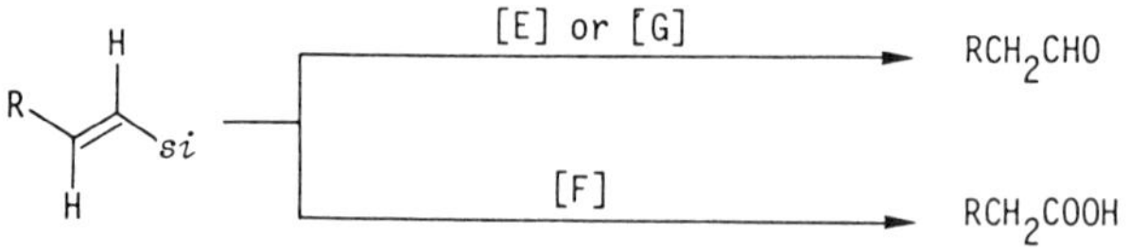

$$\xrightarrow{\text{[E] or [G]}} RCH_2CHO \qquad (9)$$

$$\xrightarrow{\text{[F]}} RCH_2COOH \qquad (10)$$

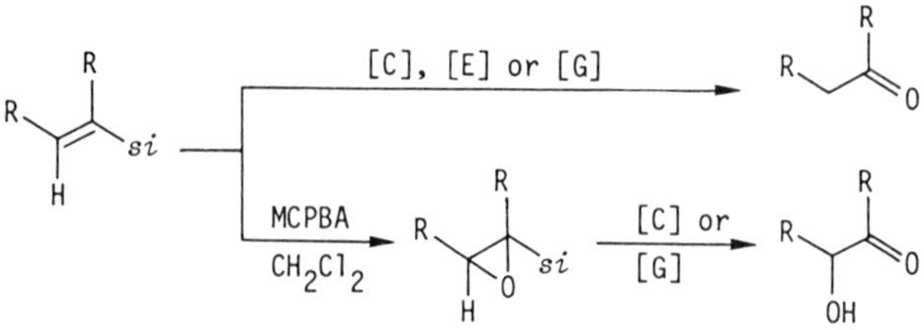

$$\xrightarrow{\text{[C], [E] or [G]}} \qquad (11)$$

$$\xrightarrow[\text{CH}_2\text{Cl}_2]{\text{MCPBA}} \qquad \xrightarrow[\text{[G]}]{\text{[C] or}} \qquad (12)$$

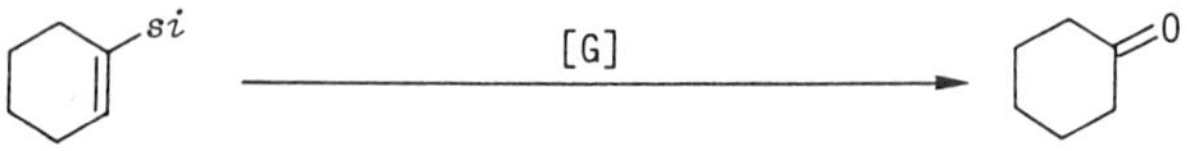

$$\xrightarrow{\text{[G]}} \qquad (13)$$

A further significant feature is the selective oxidation of an alkenyl–alkoxysilyl compound in the presence of an alkenyl–SiMe$_3$ compound (eq. 14).

$$C_6H_{13}\diagdown\diagup SiMe_3 \quad + \quad \text{(cyclohexenyl)}-SiMe(OEt)_2$$

$$\xrightarrow{[G]} \quad C_6H_{13}\diagdown\diagup SiMe_3 \quad + \quad \text{(cyclohexanone)} \qquad (14)$$

$$\text{97\% recovered} \qquad\qquad 74\%$$

C. Synthetic applications of the oxidative cleavage reactions

(a) Hydrosilylation of olefins and acetylenes

A new one-pot synthesis of anti-Markownikoff alcohols from terminal olefins has been achieved by the hydrosilylation with alkoxysilanes and the subsequent oxidation (eq. 15). Acetylenes

$$RCH=CH_2 \; + \; HSiMe(OEt)_2 \xrightarrow{\text{cat.}} RCH_2CH_2SiMe(OEt)_2 \xrightarrow{[O]} RCH_2CH_2OH \qquad (15)$$

are also converted to carbonyl compounds. Several representative results are shown in eqs. 16–21 [8,12].

$$Br(CH_2)_9CH=CH_2 \xrightarrow{a} \xrightarrow{[E]} Br(CH_2)_{11}OH \qquad\qquad 73\% \qquad (16)$$

$$MeO_2C(CH_2)_8CH=CH_2 \xrightarrow{a} \xrightarrow{[E]} MeO_2C(CH_2)_{10}OH \qquad 79\% \qquad (17)$$

$$Ph\text{-}\overset{O}{\diagdown}\diagup\diagdown\diagup \xrightarrow{b} \xrightarrow{[G]} Ph\text{-}\overset{O}{\diagdown}\diagup\diagdown\diagup\text{-}OH \qquad 73\% \qquad (18)$$

$$BuC\equiv CBu \xrightarrow{b} Bu\diagup\overset{Bu}{\diagdown}_{si}$$

$$\xrightarrow{[E]} Bu\diagdown\overset{Bu}{\underset{O}{\diagup}} \qquad 81\% \qquad (19)$$

$$\xrightarrow{c} \xrightarrow{[G]} Bu\diagdown\overset{Bu}{\underset{\underset{OH}{O}}{\diagup}} \qquad 58\% \qquad (20)$$

$$RC\equiv CH \xrightarrow{b} R\diagdown\diagup_{si} \xrightarrow{[F]} RCH_2COOH \qquad\qquad (21)$$

$$R = n\text{-}C_6H_{13}\text{-} \quad 68\%, \quad t\text{-}C_4H_9\text{-} \quad 73\%, \quad Ph\text{-} \quad 61\%^d$$

^{a} HSiMe(OEt)$_2$, RhCl(PPh$_3$)$_3$ (0.5 mol%), 80°C, 3-9 h.
^{b} HSiMe(OEt)$_2$, H$_2$PtCl$_6$ cat., r.t., 0.5 h.
^{c} MCPBA, CH$_2$Cl$_2$, 0°C, 5 h. ^{d} Hydrosilylated with HSi(OEt)$_3$.

     In view of the simplicity due to the use of commercially
available air-stable hydrosilanes, high regioselectivity, and a
wide functional group compatibility, the present methods may be
useful alternatives to the existing hydrometalation-oxidation
procedures.

(b) Intramolecular hydrosilylation

     In principle, intramolecular hydrosilylation may be
attained with a hydrosilyl group anchored to a neighboring
functional group, as shown in eq. 22. The following singnifi-
cant features are envisaged: (1) acceleration, (2) high regioselectivity, (3) high stereoselectivity, and (4) high chemoselectivity. Typical examples have been obtained in the transformation of homoallyl alcohols into 1,3-diols and of homo-
propargyl alcohols into β-hydroxyketones (eqs. 23-28). In all
cases, $(HMe_2Si)_2NH$ can conveniently be used for protection of
the hydroxy group in the first step [14].

$$ \text{(22)} $$

$$ \text{(23)} \quad 69\% $$

$$ \text{(24)} $$

$$ \text{(25)} $$

From E 36 / 64   total 82%
From Z 45 / 55   total 74%

$$ \text{(26)} \quad 73\% $$

$$ \text{(27)} \quad 50\% $$

$$ \text{(28)} \quad 65\% $$

a $(HMe_2Si)_2NH$. b H_2PtCl_6, 60°C, 10 min. c $RhCl(PPh_3)_3$, 100°C, 20 min.

Intramolecular hydrosilylation thus provides a useful methodology to overcome two big drawbacks in the intermolecular hydrosilylation of olefins and acetylenes: i.e., almost inertness of internal olefins and low regioselectivity encountered with unsymmetrical internal acetylenes.

(c) Nucleophilic hydroxymethylation with silafunctional silylmethyl Grignard reagents

The following three silafunctional silylmethyl Grignard reagents have been found to be new practically useful nucleophilic hydroxymethylating agents towards a variety of electrophiles (eq. 29), such as organic halides [10], tosylates [15], epoxides [15], aldehydes [16], ketones [16], and α,β–unsaturated ketones [16] (eqs. 30–39).

$$si\text{-}CH_2MgCl \quad + \quad E^{(+)} \longrightarrow si\text{-}CH_2\text{-}E \longrightarrow HOCH_2\text{-}E \qquad (29)$$

$(i\text{-}PrO)_2MeSiCH_2MgCl$ [I]

$(i\text{-}PrO)Me_2SiCH_2MgCl$ [II] $\equiv$ $HOCH_2^{(-)}$

$(\diagdown\diagup)Me_2SiCH_2MgCl$ [III]

$$n\text{-}C_8H_{17}Br \xrightarrow{a} n\text{-}C_8H_{17}CH_2\text{-}si \xrightarrow{[E]} n\text{-}C_8H_{17}CH_2OH \qquad 81\% \qquad (30)$$

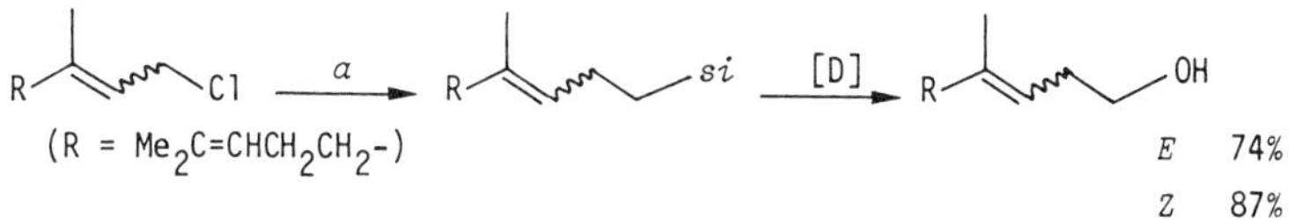

(R = $Me_2C=CHCH_2CH_2$-)

E   74%

Z   87%   (31)

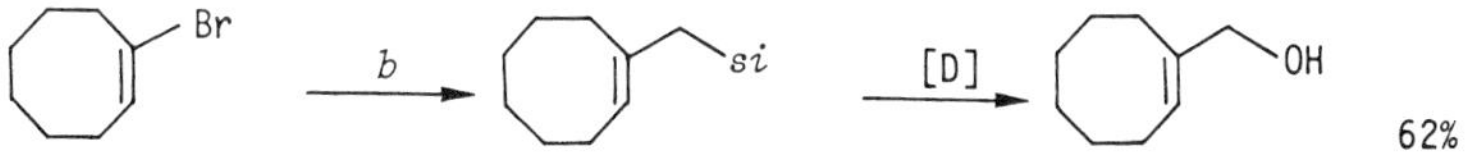

62%   (32)

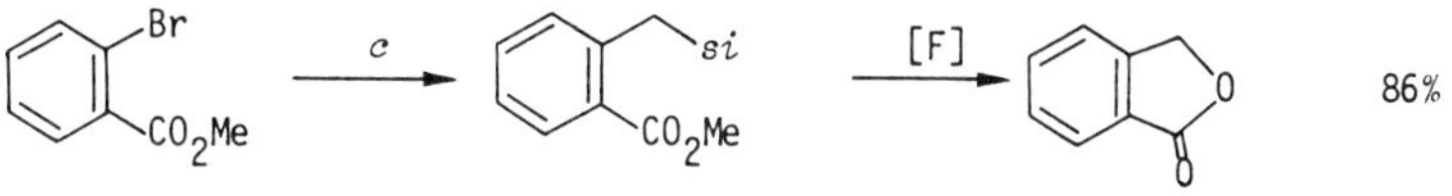

86%   (33)

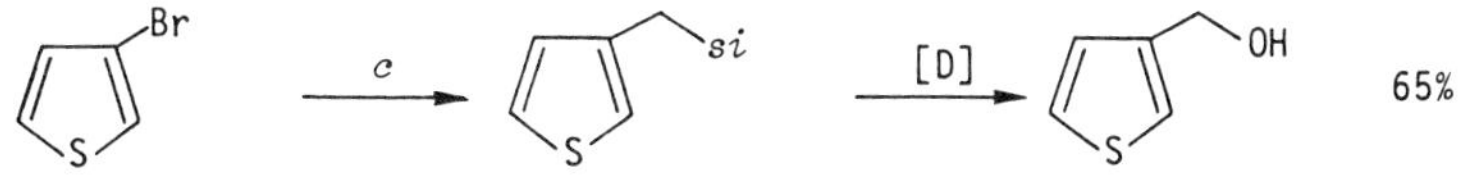

65%   (34)

$$n\text{-}C_8H_{17}OTs \xrightarrow{d} n\text{-}C_8H_{17}CH_2\text{-}si \xrightarrow{[G]} n\text{-}C_8H_{17}CH_2OH \qquad 67\% \qquad (35)$$

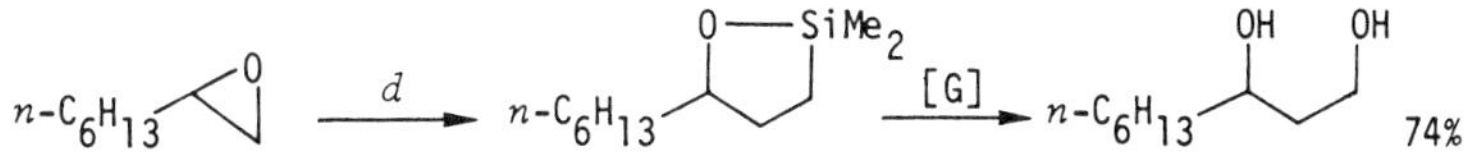

74%   (36)

$$n\text{-}C_6H_{13}\text{-}C(=O)\text{-}CH_3 \xrightarrow{e} n\text{-}C_6H_{13}\text{-}C(OH)(CH_2\text{-}si) \xrightarrow{[G]} n\text{-}C_6H_{13}\text{-}C(OH)(CH_2OH) \quad 86\% \quad (37)$$

$$\underset{n}{\bigcirc}\!=\!O \xrightarrow{e} \underset{n}{\bigcirc}\!(HO)(si) \xrightarrow{[G]} \underset{n}{\bigcirc}\!(HO)(OH) \quad (38)$$

$$n = 5,\ 6,\ 8,\ 12 \quad\quad 46\text{-}86\%$$

$$\xrightarrow{f} \text{(}si\text{)} \xrightarrow{g,\ [G]} \text{(}CH_2OH\text{)} \quad 68\% \quad (39)$$

[a] [I], CuI (10 mol%), THF.   [b] [I], NiCl$_2$(dppp) (1 mol%), THF.
[c] [I]/ZnCl$_2$, PdCl$_2$(dppf) (1 mol%), THF.   [d] [II], CuI (10 mol%), THF.
[e] [II], THF, 0°C.   [f] [III], CuI (10 mol%), Et$_2$O.   [g] CF$_3$COOH, KHF$_2$, 50°C.

(d) Asymmetric synthesis via chiral silafunctional compounds: chiral α-hydroxyalkyl anion equivalents

Addition of Grignard or organolithium reagents to sila-functional vinylsilanes provides useful routes to carbanions α to silicon. Trapping with electrophiles and the subsequent oxidation afford the corresponding alcohols in which the sila-functional vinylsilanes act as a new two-carbon building block and in turn the α-silyl carbanion as an α-hydroxyalkyl anion equivalent (eq. 40) [17,18].

$$R\text{-}m \ +\ \overset{si}{/}\!\!\!= \longrightarrow R\!\!\underset{m}{\overset{si}{\diagdown\!\diagup}} \longrightarrow R\!\!\underset{E}{\overset{si}{\diagdown\!\diagup}} \longrightarrow R\!\!\underset{E}{\overset{OH}{\diagdown\!\diagup}} \quad (40)$$

Chiral vinylsilanes containing optically active functional groups on silicon have been used as precursors for chiral α-hydroxyalkyl anion equivalents. A typical transformation is shown in eq. 41 [19].

$$n\text{-BuLi} \ +\ \overset{SiMe(-N(Me)(OMe)CHPh)_2}{/}\!\!\!= \xrightarrow{a} \cdots \text{SiMe}(-N(Me)(OMe)CHPh)_2,\ MgBr$$

$$\xrightarrow{b} \overset{(-SiMe\text{-}O\text{-})_n}{\diagdown\!\diagup} \xrightarrow{[D]} \overset{OH}{\diagdown\!\diagup}\!\!* \quad (41)$$

$$60\%\ ee\ (R)$$

[a] 1) Et$_2$O, 0°C, 2) MgBr$_2$.
[b] 1) CH$_2$=CHCH$_2$Br, CuI (10 mol%), THF, r.t., 2) 6N HCl.

(e) Miscellaneous studies

The following several subjects also have dealt with the oxidative cleavage reactions of silafunctional compounds. They include (1) synthetic applications of intramolecular rearrangement of chloromethyl(alkenyl)silanes [20], (2) conversion of organometallics to alcohols via silylation and oxidation [21], (3) asymmetric synthesis of optically active alcohols via the catalytic asymmetric hydrosilylation [22,23], (4) determination of absolute configuration of optically active organosilicon compounds [24], (5) oxidation of alkylsilatranes [25], (6) intramolecular hydroxymethylation of allylic alcohols [26], and (7) utility of phenyldimethylsilylcuprate as a hydroxide synthon [27].

CONCLUSION

I have described several synthetically useful processes, all of which are accomplished only with silafunctional compounds. Even these few results suggest that silafunctional organosilicon chemistry will provide a new area in synthetic organic chemistry for future consideration. We have just reached this new field.

ACKNOWLEDGEMENT

It is a pleasure to acknowledge the significant contributions of the coworkers, whose names are recorded in the references. My sincere thanks are also due to Professor Emeritus Makoto Kumada for encouragement.

REFERENCES

1. (a) Colvin, E. "Silicon in Organic Synthesis", Butterworths, London, 1981. (b) Weber, W. P. "Silicon Reagents for Organic Synthesis", Springer-Verlag, Berlin, 1983.
2. Corriu, R. J. P.; Guerin, C. Adv. Organomet. Chem. **1982**, 20, 265.
3. Kumada, M.; Tamao, K.; Yoshida, J. J. Organomet. Chem. **1982**, 239, 115.
4. Tamao, K.; M. Akita, K.; Kumada, M., unpublished results.
5. Tamao, K.; Maeda, K., unpublished results.
6. Tamao, K.; Kakui, T.; Iwahara, T.; Kanatani, R.; Yoshida, J.; Kumada, M. Tetrahedron **1983**, 39, 983.
7. Buncel, E.; Davies, A. G. J. Chem. Soc. 1958, 1550.
8. Tamao, K.; Ishida, N.; Tanaka, T.; Kumada, M. Organometallics **1983**, 2, 1694.
9. Tamao, K.; Ishida, N. J. Organomet. Chem. 1984, 269, C37.

10. Tamao, K.; Ishida, N.; Kumada, M. J. Org. Chem. 1983, 48, 2120.

11. Tamao, K.; Akita, M.; Kumada, M. J. Organomet. Chem. 1983, 254, 13.

12. Tamao, K.; Kumada, M.; Maeda, K. Tetrahedron Lett. 1984, 25, 321.

13. (a) Stork, G,; Colvin, E. J. Am. Chem. Soc. 1971, 93, 2080. (b) Hudrlik, P. F.; Arcoleo, J. P.; Schwartz, R. H.; Misra, R. N.; Rona, R. J. Tetrahedron Lett. 1977, 591.

14. Tamao, K.; Nakajima, T.; Tanaka, T., unpublished results.

15. Tamao, K.; Ishida, N., unpublished results.

16. Tamao, K.; Ishida, N. Tetrahedron Lett., in press.

17. Tamao, K.; Kanatani, R.; Kumada, M. Tetrahedron Lett. 1984, 25, 1905.

18. Tamao, K.; Iwahara, T.; Kanatani, R.; Kumada, M. Tetrahedron Lett. 1984, 25, 1909.

19. Tamao, K.; Kanatani, R.; Kumada, M. Tetrahedron Lett. 1984, 25, 1913.

20. Tamao, K.; Nakajima,; Kumada, M. Organometallics, in press.

21. Tamao, K.; Kanatani, R., unpublished results.

22. Hayashi, T.; Tamao, K.; Katsuro, Y.; Nakae, I.; Kumada, M. Tetrahedron Lett. 1980, 21, 1871.

23. Hayashi, T.; Kabeta, K.; Yamamoto, T.; Tamao, K.; Kumada, M. Tetrahedron Lett. 1983, 24, 5661.

24. Hayashi, T.; Okamoto, Y.; Kumada, M. J. Chem. Soc., Chem. Commun. 1982, 1072.

25. Hosomi, A.; Iijima, S.; Sakurai, H. Chem. Lett. 1981, 243.

26. Nishiyama, H.; Kitajima, T.; Matsumoto, M.; Itoh, K. J. Org. Chem. 1984, 49, 2298.

27. Fleming, I.; Henning, R.; Plaut, H. J. Chem. Soc., Chem. Commun. 1984, 29.

CHAPTER 22

THE STRUCTURE-REACTIVITY CORRELATION IN HETARYLSILANES

E. Lukevics Institute of Organic Synthesis, Latvian Academy of Sciences, Aizkraukles 21, Riga 226006, U.S.S.R.

INTRODUCTION

The electronacceptor character of silyl groups bound to a furan or thiophene ring was demonstrated by various physico-chemical methods: ^{1}H, ^{13}C, ^{29}Si NMR, ESR and photoelectronic spectroscopy.

This fact was interpreted in terms of p_{π}-d_{π}-interaction between the π-electrons of the heterocycle and the vacant 3d-orbitals of the silicon atom [1-3].

In order to study the influence of this effect on the reactivity of the Si-H bond in furyl- and thienylhydrosilanes, we have synthesized a number of hetarylsilanes I-VI and investigated their reactivity in the reactions of hydrosilylation of acetylene derivatives and dehydrocondensation with hydroxyl-containing compounds.

RESULTS AND DISCUSSION

The hydrosilanes I-III, V were obtained in the reaction between 2- and 3-furyl-, thienyl- and 2-(4,5-dihydrofuryl)lithium and the corresponding chlorosilanes; IV - by the reduction of furylethoxysilanes and thienylchlorosilanes with lithium aluminium hydride; and 2-tetrahydrofurylsilanes VI were synthesized by hydrogenation of 2-(4,5-dihydrofuryl)silanes V at 25°C in the presence of 5% Pd/Al$_2$O$_3$.

Butyllithium obtained by ultrasonic irradiation of the heterogeneous system Li/n-BuLi-hexane at room temperature was used for hetaryllithium synthesis. The reactions with chlorohydrosilanes were carried out at -30 - -50°C, but even in these conditions a partial (6-45%) substitution of the Si-H bond by the hetaryl group occured.

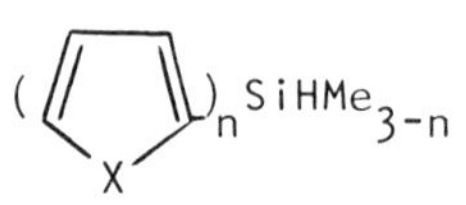

I

X = 0, S; n = 1-3

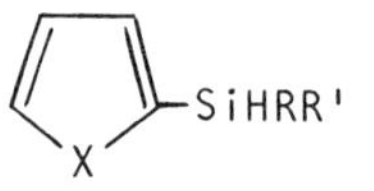

II

X = 0, S;
R,R'=alkyl, aryl

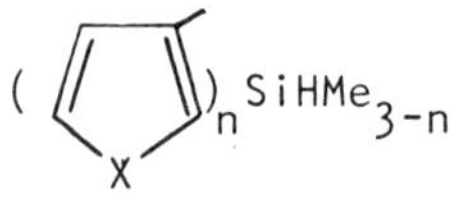

III

X = 0, S; n = 1-3

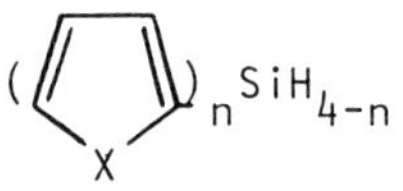

IV

X = 0, S; n = 1-3

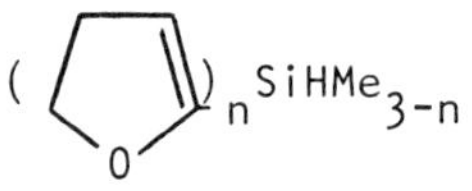

V

n = 1-3

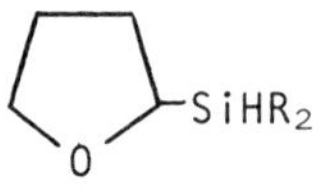

VI

R = alkyl

During the synthesis of 2-tetrahydrofurylsilanes VI, after the completion of hydrogenation or due to an increase in substrate concentration a rearrangement to 2,2-dialkyl-1-oxa-2-sila-cyclohexane (VII) readily takes place under the influence of the same palladium catalyst [4]:

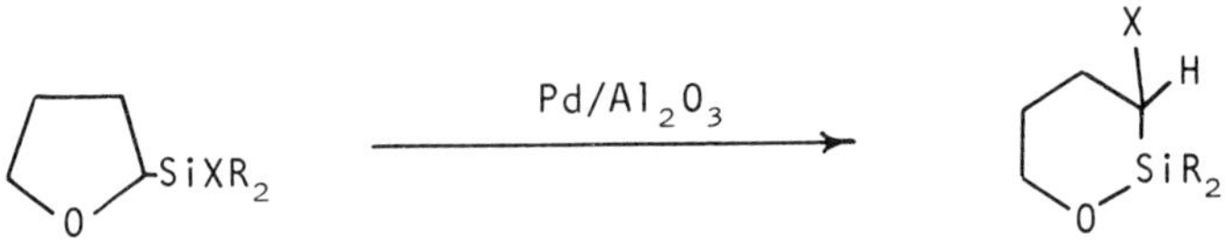

X = H, D

VII

In most cases, the hydrosilylation of acetylene derivatives with hydrosilanes I, II, V, VI in the presence of chloroplatinic acid yields β-trans and α-isomers:

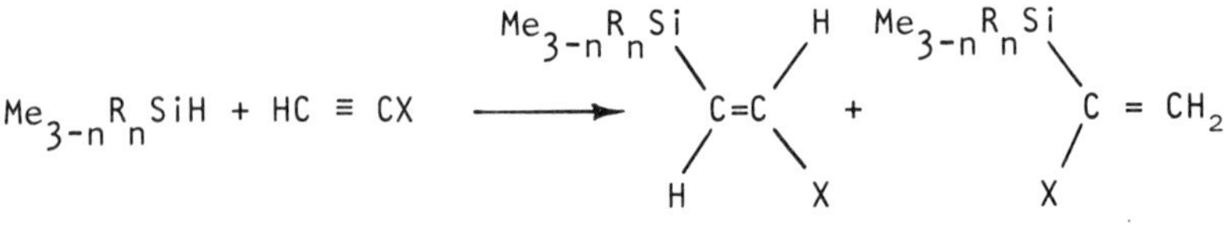

R =

X = Ad, Ph,

The isomer ratio depends essentially on the structure of
the acetylene derivatives and, to a smaller degree, on the structure of the hydrosilane. Thus, the hydrosilylation of 1-ethynyl-
adamantane (50°C, 6 hrs) with thienylhydrosilanes gives only the
β-trans isomer (yield ~90%). In the case of phenylacetylene a
mixture of isomers is obtained (67-89%) where the β-isomer (70-
75%) prevails, while in the reaction with 2-ethynylthiophene
(20-50%) the α-isomer is predominant (~70%). 2-Propargylthio-
phene reacts with thienylhydrosilanes much easier than 2-ethy-
nyltiophene. Thereby, the adducts are obtained (80-98%) with the
β-trans isomer prevailing. During the hydrosilylation of phenylacetylene with the furylhydrosilanes I, V and VI (100°C, 0.5
h) the β-isomer content increases from tetrahydrofuryl- to di-
hydrofuryl- and furylsilanes (from 70 to 92%).

The α-isomer is chiefly formed (55-68%) in the hydrosilylation of methyl propiolate with the silanes I, II. Dimethyl(2-
thienyl)silane reacts quantitatively already at 50°C. The yields
of di- and trihetaryl adducts are not large in these conditions
and reaches 95% only when the reaction temperature is increased
to 150°C. By increasing the number of hetaryl groups in the silane molecule the β-isomer content increases from 30 to 45% in
the reaction products.

N-Propargylamines undergo hydrosilylation less readily and
it is necessary to raise the temperature of the reaction to 150-
180°C. The yield of the reaction products become smaller with
increasing number of hetaryl groups in the silanes I-III. In the
case of compounds IV, on the contrary, the yield of hydrosilylation falls when the number of Si-H bonds is increased. The amount
of the β-trans isomer grows with the number of hetaryl groups
at the silicon atom in compounds I-III while in the reaction of
tri(2-thienyl)silane and N-propargylperhydroazepine the β-trans
isomer is obtained exclusively.

The yields of the hydrosilylation of propargylamines with
dimethyl(2-thienyl)silane depend on the catalyst used and dec-
line in the sequence:

$$H_2PtCl_6 \cdot 6H_2O > (Ph_3P)_3RhCl > H_2OsCl_6 \cdot 6H_2O > Co_2(CO)_8$$

The dehydrocondensation (20%) at the hydroxyl group proceeds alongside with hydrosilylation ($\alpha:\beta$ = 39:41) when propargyl alcohol reacts with dimethyl(2-thienyl)silane (50°C, 6 hrs) in the presence of $H_2PtCl_6 \cdot 6H_2O$:

$$Me_2RSiH + HC \equiv CCH_2OH \xrightarrow{H_2PtCl_6} Me_2RSi \diagdown C = C \diagup H \quad H \diagup C = C \diagdown CH_2OH \quad +$$

$$+ \quad Me_2RSi \diagdown C = CH_2 \diagup HOCH_2 \quad + \quad Me_2RSi \diagdown C = C \diagup H \quad H \diagdown CH_2OSiRMe_2 \quad + \quad Me_2RSi \diagdown C = CH_2 \diagup Me_2RSiOCH_2$$

R = 2-thienyl

In the case of di- and trithienylsilanes only the hydrosilylation of the triple bond occurs. The regioselectivity increases with the number of thienyl groups in the silane molecule.

The silyl group attacks mainly the site of the highest electronic density during the hydrosilylation of acetylene derivatives with hetarylsilanes. The increase in the β-isomer amount with the number of hetaryl groups irrespective of the direction of triple bond polarization can be accounted for by the steric hindrance during the formation of the α-isomer. Owing to this, the silyl group is directed to the terminal carbon atom. A reduction in the reaction rate observed in the same sequence can be explained by the influence of the p_π-d_π-interaction, since the growth of the total inductive effect (-I) of the substituents at the silicon atom would have accelerated the reaction. This can be seen in the hydrocondensation of hetarylsilanes I-III, V with hydroxyl-containing compounds in the presence of amines:

$$RR_2'SiH + HOR'' \xrightarrow{\text{amine}} RR_2'SiOR'' + H_2$$

R = alkyl, aryl, thienyl, furyl, alkoxy

R' = aryl, thienyl, furyl, dihydrofuryl, alkoxy

R'' = alkyl, aryl, -N=CRR', OCR.

The reactivity of alkanols, phenols and oximes in the reaction of dehydrocondensation increases with the -I-effect of substituents. The logarithms of reaction rates are linearly correlated with the inductive Taft's constants of the substituents. The reaction of dehydrocondensation with carboxylic acids is retarded by the electronacceptor substituents in the acid molecule.

The logarithms of the reaction rate constants for aromatic acids correlate linearly with the Hammett constants of the substituents.

The rate of dehydrocondensation of the hetarylsilanes I-III, V depends on the radicals at the silicon atom and drops in the sequence:

The reaction rate and product yields increase with the number of hetaryl groups in the silane molecule.

The logarithms of the reaction rate constants in the reaction between the hetarylsilanes I-III and propargyl alcohol in the presence of piperidine at 33°C correlate with spin-spin coupling constants $^1J_{SiH}$:

$$lgk = -33.5 + 0.17\,^1J_{SiH} \qquad r = 0.975$$

and with the sum of Taft's constants of the substituents at the silicon atom in the hydrosilanes [5]:

$$lgk = -2.72 + 1.37\,\Sigma\sigma^*_R \qquad r = 0.982$$

The reactivity of hydrosilanes in the reaction with phenol, acetophenone oxime, and acetic acid in the presence of triethylamine also increases with growing electronacceptor properties of the substituents at the silicon atom.

The kinetic data obtained indicate that the dehydrocondensation occurs by the molecular mechanism through an intermediate complex containing a pentacoordinated silicon atom. Its formation is facilitated by the electronacceptor substituents in the silane molecule, which by increase the positive charge on the silicon atom promotes the Si ... O coordination. The latter prevails over the $p_\pi - d_\pi$-interaction of the hetaryl substituents and the silicon atom.

Hence, if the reactivity of hetarylsilanes in the reaction of hydrosilylation can be interpreted in terms of p_π-d_π-interaction slowing down the reaction; their alcoholysis rate increases with the -I-effect of the substituents at the silicon atom, the effect of p_π-d_π-interaction being obscured.

REFERENCES

1. Lukevics, E., Erchak, N.P. In: Advances in Furan Chemistry,

Riga, 1978, p. 198 (in Russian).
2. Lukevics, E., Skorova, A.E., Pudova, O.A. Thiophene derivatives of group IVB elements. Sulfur Reports, 1982, 2, 177.
3. Zykov, B.G., Erchak, N.P., Khvostenko, V.I., Lukevics, E., Matorikina, V.F., Asfandiarov, N.I. Photoelectron spectra of furylsilanes and their carbon analogs. J. Organomet. Chem., 1983, 253, 301.
4. Lukevics, E., Gevorgyan, V., Goldberg, Y., Popelis, J., Gavars, M., Gaukhman, A., Shimanska, M. Synthesis of 2-(4,5-dihydrofuryl)- and 2-(tetrahydrofuryl)dimethylhydrosilanes. Rearrangement of 2-(tetrahydrofuryl)dimethylhydrosilane with ring expansion. Heterocycles, 1984, 22, 987.
5. Lukevics, E., Dzintara, M. Silylation of hydroxyl-containing compounds with aryl- and heteroarylhydrosilanes in the presence of amines. J. Organomet. Chem., 1984 (in press).

PART 5

BIOORGANOSILICON CHEMISTRY AND FURTHER APPLICATIONS

RECENT RESULTS IN BIOORGANOSILICON CHEMISTRY: NOVEL SILA-DRUGS AND MICROBIAL TRANSFORMATIONS OF ORGANOSILICON COMPOUNDS

Reinhold Tacke Institute of Inorganic and Analytical Chemistry, Technical University, D–3300 Braunschweig (F.R.G.)

NOVEL SILA-DRUGS

In the course of our systematic studies on C/Si-bio-isosterism a large number of sila-substituted drugs (sila-drugs) belonging to various types of structures have been synthesized and investigated pharmacologically in comparison with their corresponding carbon analogues (Fig. 1).

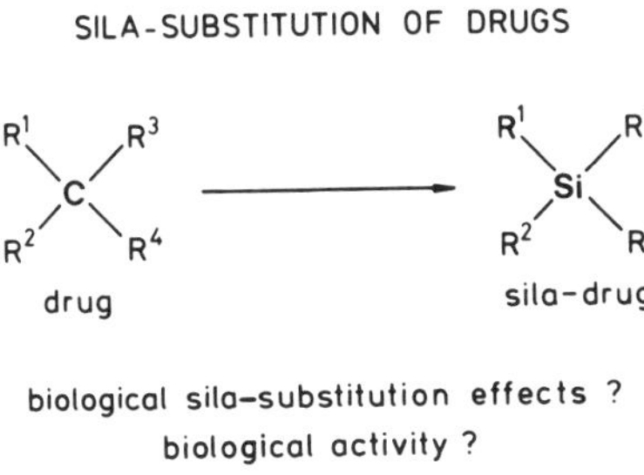

SILA-SUBSTITUTION OF DRUGS

biological sila–substitution effects ?
biological activity ?

Figure 1

The general types of carbon/silicon pairs investigated
so far for biological sila-subtitution effects are
shown in Fig. 2. In this paper some recent results
concerning the sila-substitution of biologically
active carbinols will be presented.

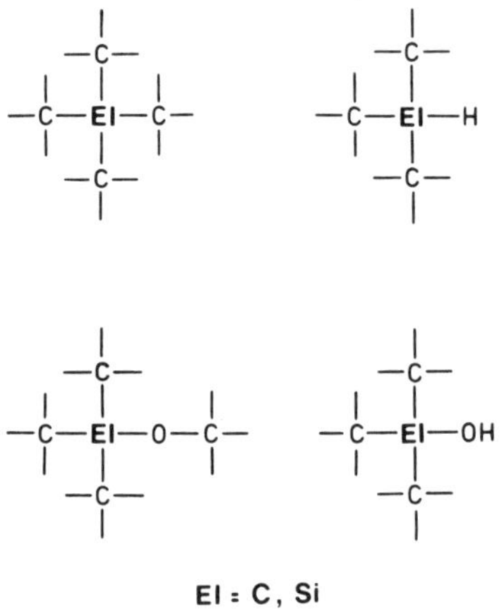

Figure 2

EI : C, Si

In continuation of our studies /1/ on sila-substituted
antimuscarinic agents we have synthesized the sila-
drugs sila-procyclidine, sila-trihexyphenidyl, sila-
cycrimine, sila-tricyclamol iodide, sila-tridihexethyl
chloride, and sila-hexocyclium methylsulfate /2,3/
(Fig. 3). The synthetic approach to these compounds is
demonstrated in Fig. 4 /2/ and Fig. 5 /4/ showing two
different syntheses of sila-procyclidine. All the
silanols shown in Fig. 3 and their carbon analogues,
the latter being used therapeutically as spasmolytics
(tricyclamol chloride, tridihexethyl chloride, hexo-
cyclium methylsulfate) and as antiparkinsonian drugs

(procyclidine, trihexyphenidyl, cycrimine), were
tested comparatively for their antimuscarinic properties in guinea-pig ileum using acetylcholine as agonist /2,3/. The silanols were found to be potent antimuscarinic agents, and most of them exhibit a greater
antimuscarinic potency than their corresponding carbon
analogues. As the hydroxy group of these drugs seems
to play an important role in the interaction with the
muscarinic receptors /1/, the greater OH-acidity of
the silanols relative to that of the carbinols may be
responsible for this effect.

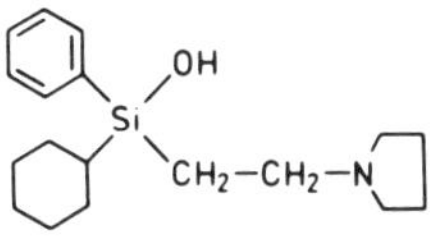

sila-procyclidine

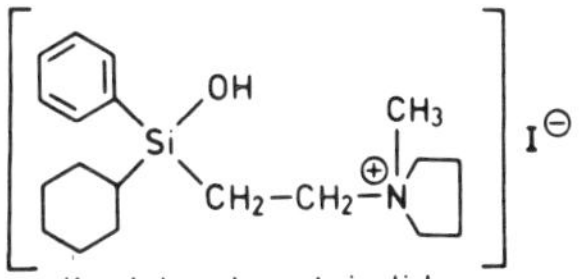

sila-tricyclamol iodide

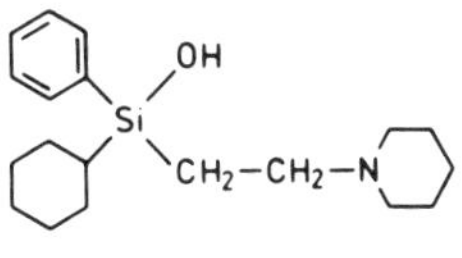

sila-trihexyphenidyl

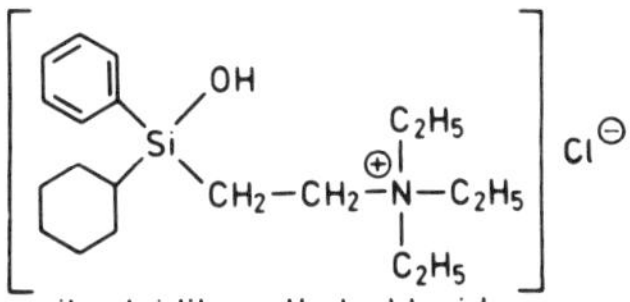

sila-tridihexethyl chloride

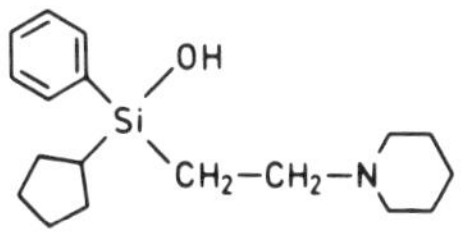

sila-cycrimine

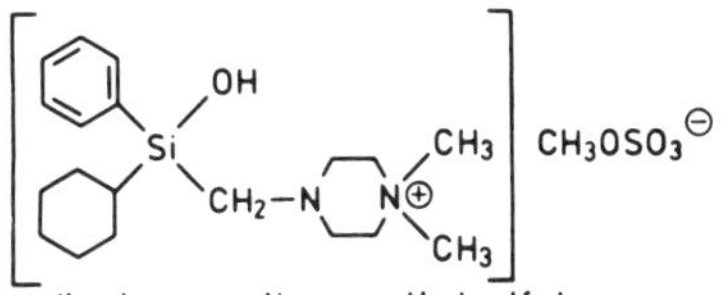

sila-hexocyclium methylsulfate

Figure 3

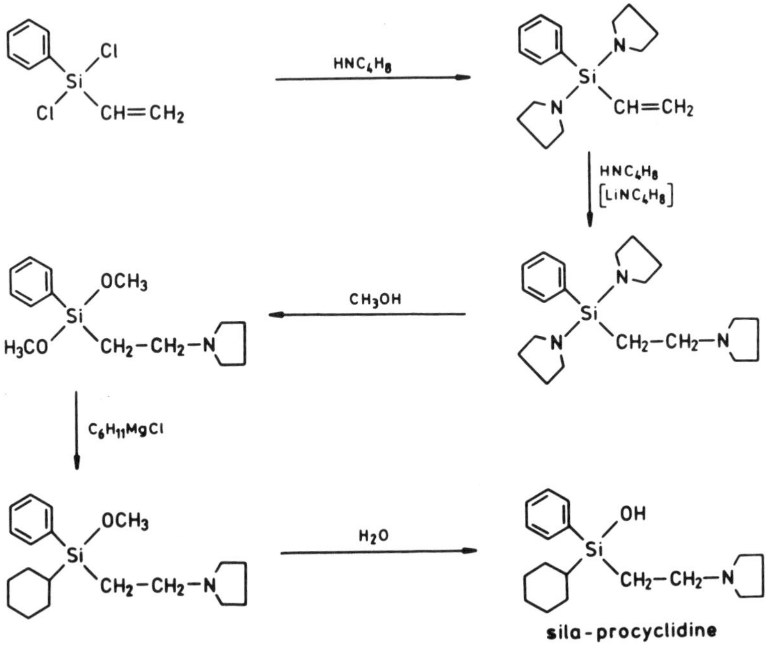

Figure 4

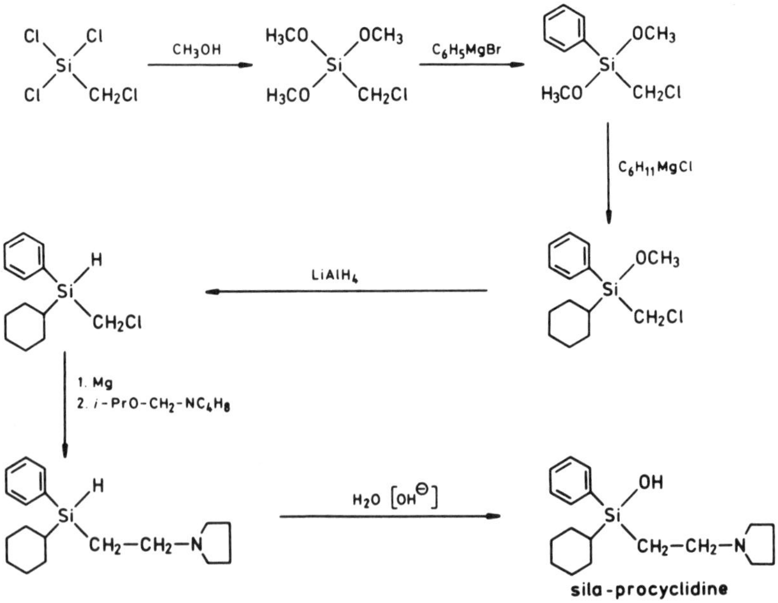

Figure 5

The sila-drugs shown in Fig. 3 and their carbon analogues possess a chiral centre, and the pharmacological experiments mentioned above were first carried out
with the racemates. However, from the literature it
was known that some of the carbon compounds exhibit a
high stereoselectivity at muscarinic receptors. For
example, the (R)-configurated enantiomers of both
procyclidine and tricyclamol iodide are much more
potent than the corresponding (S)-configurated enantiomers (Fig. 6). The (R)-enantiomer is 375 and 87 times,
respectively, more potent than the (S)-enantiomer.
For this reason, we were highly interested in the
stereoselectivity of the corresponding silicon analogues at muscarinic receptors, and thus prepared the
enantiomers of sila-procyclidine (resolving agents:
L(+)- and D(-)-tartaric acid, respectively) and sila-

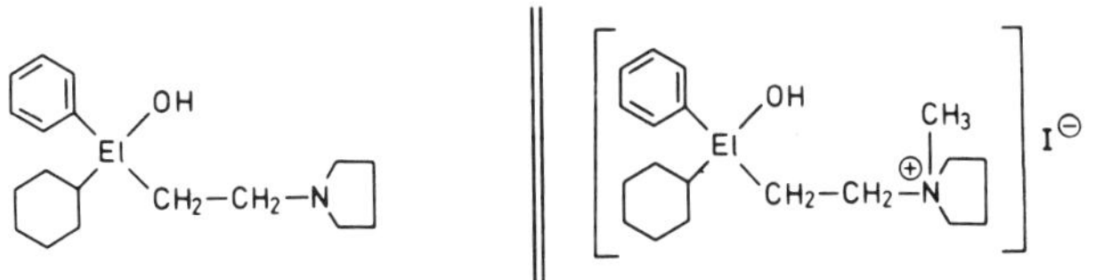

| El = C : procyclidine | El = C : tricyclamol iodide |
| El = Si : sila-procyclidine | El = Si : sila-tricyclamol iodide |

relative antimuscarinic potencies (guinea-pig ileum)[*]

| El = C | R : S | 375 : 1 | El = C | R : S | 87 : 1 |
| El = Si | R : S | ? | El = Si | R : S | ? |

[*]R.B. Barlow, J. Pharm. Pharmacol. 23, 90 (1971).

Figure 6

tricyclamol iodide as shown in Fig. 7 /5/. The enantiomeric excess of the samples obtained was found to be
> 96% (^{13}C nmr studies with Eu(hfc)$_3$ as shift reagent).
The absolute configuration of the enantiomers was determined on the basis of the X-ray structural analysis
of the laevorotatory enantiomer of sila-procyclidine.
With respect to the projected pharmacological studies,
the racemisation stability of the silanols in aqueous
solution was found to be sufficient.

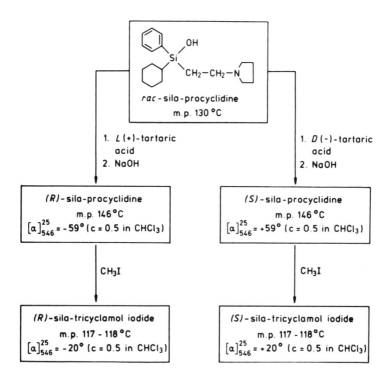

Figure 7

The enantiomers of procyclidine, sila-procyclidine, tricyclamol iodide, and sila-tricyclamol iodide were investigated comparatively on the isolated guinea-pig ileum using carbachol as agonist /6/. By analogy to the (R)-enantiomers of the carbon compounds, the (R)-configurated sila-analogues were found to have a greater antimuscarinic potency than the corresponding (S)-enantiomers. However, the stereoselectivity index of the carbon compounds is much more pronounced. This is shown in Fig. 8 by the pA_2-values (a measure of affinity for muscarinic receptors) of procyclidine and sila-procyclidine. In contrast to the high index of 375 for the enantiomers of procyclidine, a value of about 10 was found for the enantiomers of the sila-analogue. Whereas (R)-procyclidine and (R)-sila-procyclidine have approximately the same high anti-muscarinic potency, (S)-sila-procyclidine was found to

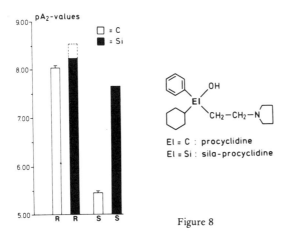

El = C : procyclidine
El = Si : sila-procyclidine

Figure 8

be about two orders of magnitude more potent than the (S)-configurated carbon analogue. In comparison with these results, the difference in the stereoselectivity indices of tricyclamol iodide and sila-tricyclamol iodide was found to be smaller.

In the course of the studies described above, various analogues of procyclidine and sila-procyclidine have also been investigated for their affinities to cardiac muscarinic receptors /1,2,7/. All the compounds tested have been found to exhibit a greater affinity for ileal than for atrial muscarinic receptors of the guinea pig. As a rule, the degree of this pharmacological selectivity is more pronounced for the silicon compounds than for their carbon analogues.

In continuation of these studies cyclohexylphenyl-(3-piperidinopropyl)silanol, inter alia, was synthesized (Fig. 9) and investigated pharmacologically /7,8/. This compound (tested as racemate) was found to exhibit a high degree of selectivity for muscarinic receptors of the ileum, as reflected by a K_D-ratio (atria/ileum) of 27 (Fig. 9). The selectivity of the corresponding carbon analogues is significantly less pronounced (K_D-ratio = 16). To our knowledge, cyclohexylphenyl(3-piperidinopropyl)silanol is the best antagonistic tool known today to differentiate between cardiac and ileal muscarinic receptors.

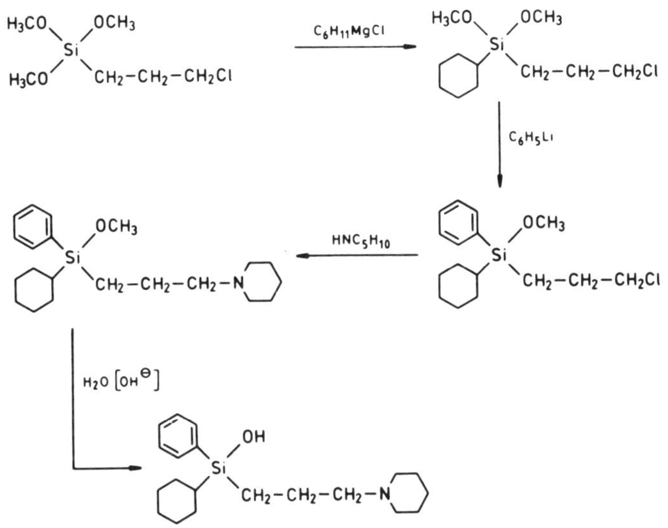

antimuscarinic potencies at muscarinic receptors
of guinea-pig ileum and atria [a]

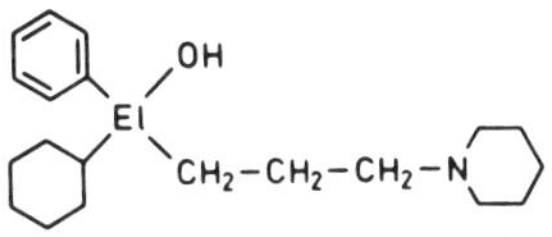

	El	pA_2		K_D-ratio [b]
		ileum	atria(force)	(atria/ileum)
	C	7.98	6.71	19
	Si	7.96	6.53	27

a) agonist: arecaidine propargyl ester
b) $pA_2 = -\log K_D$

Figure 9

From the studies presented in this chapter, it may be
concluded: As a rule, sila-substitution leads to
organosilicon compounds still having the same mode of
pharmacological action (the pharmacodynamics are not
changed qualitatively). However, the degree of pharmacological potency and the pharmacological selectivity

may be influenced substantially by a carbon/silicon
exchange. It is concluded that sila-substitution of
drugs may be an useful and efficient tool in drug
design.

MICROBIAL TRANSFORMATIONS OF ORGANOSILICON COMPOUNDS

Recently, we have reported on the microbial enantioselective reduction of the keto group of some (trimethylsilyl)methyl acetoacetates /9/. To our knowledge,
this was the first example of a systematic biotrans-
formation of organosilicon compounds.

In continuation of these studies we investigated
further transformations of silicon containing substrates, especially those of compounds with functional
groups bound directly to silicon /10/. An example is
given in Fig. 10 showing the synthesis of (1R,2S)-
1-methyl-1-phenyl-1-silacyclohexan-2-ol. Starting from
racemic 1-methyl-1-phenyl-1-silacyclohexan-2-one, this
optically active silane could be obtained by two bio-
transformation steps and one chemical step: The first
step involves the diastereoselective reduction of the
carbonyl group bound to silicon by Kloeckera corticis
to a 1:1-mixture of (1S,2R)- and (1R,2S)-1-methyl-1-
phenyl-1-silacyclohexan-2-ol (yield 97%, diastereomeric excess ∼90%). The esterefication of this product
with acetyl chloride and pyridine leads in nearly

quantitative yield to the corresponding acetate (1:1-mixture), which was subsequently transformed into the optically active alcohol with (R)-configuration at silicon and (S)-configuration at carbon by a highly enantioselective hydrolysis with Pichia pijperi. After separation from the non-reacted (1S,2R)-configurated acetate by chromatography (silica gel) the produced (1R,2S)-configurated alcohol was obtained in a very pure form (enantiomeric excess > 96%).

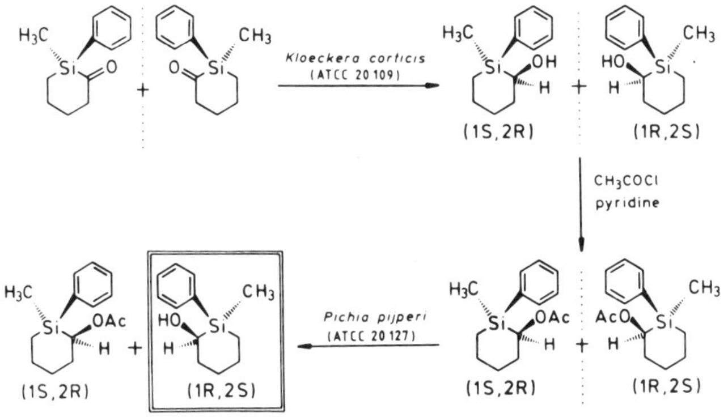

Figure 10

These results point out the high synthetic potential of microbial transformations for organosilicon chemistry, especially for the synthesis of optically active silicon compounds. Using the methods of modern biotechnology, a scale-up of such biotransformations and thus the technical synthesis of larger amounts of optically active silanes is in principle possible.

REFERENCES

/1/ R. Tacke, M. Strecker, G. Lambrecht, U. Moser and
E. Mutschler, Arch. Pharm. (Weinheim) 317, 207
(1984); and references therein.

/2/ R. Tacke, M. Strecker, G. Lambrecht, U. Moser and
E. Mutschler, Liebigs Ann. Chem. 1983, 922.

/3/ G. Lambrecht, H. Linoh, U. Moser, E. Mutschler and
R. Tacke, unpublished results.

/4/ R. Tacke and J. Pikies, unpublished results.

/5/ R. Tacke, H. Linoh, W. S. Sheldrick, L. Ernst,
G. Lambrecht, U. Moser and E. Mutschler, Erster
Gesamtkongreß der Pharmazeutischen Wissenschaften,
Abstracts, pp. 129-130, München (1983); and un-
published results.

/6/ G. Lambrecht, U. Moser, E. Mutschler, H. Linoh
and R. Tacke, Erster Gesamtkongreß der Pharmazeutischen Wissenschaften, Abstracts, pp. 179-180,
München (1983); and unpublished results.

/7/ G. Lambrecht, U. Moser, E. Mutschler, J. Wess,
H. Linoh, M. Strecker and R. Tacke, Naunyn-
Schmiedeberg's Arch. Pharmacol. 325, Suppl. R62
(1984); and unpublished results.

/8/ G. Lambrecht, H. Linoh, U. Moser, E. Mutschler,
M. Strecker, R. Tacke and J. Wess, VIIIth Inter-
national Symposium on Medicinal Chemistry,
Abstracts, p. 123, Uppsala (1984).

/9/ R. Tacke, H. Linoh, B. Stumpf, W.-R. Abraham,
K. Kieslich and L. Ernst, Z. Naturforsch. 38b,
616 (1983).

/10/ R. Tacke, H. Zilch, B. Stumpf, L. Ernst and
D. Schomburg, unpublished results.

CHAPTER 24

SiH$_4$ AND Si$_2$H$_6$ PLASMA – DIAGNOSTICS AND FILM DEPOSITION

K. Tanaka, A. Matsuda and N. Hata Electrotechnical Laboratory, 1-1-4 Umezono, Sakuramura, Niihari-gun, Ibaraki 305, Japan

INTRODUCTION

Since the initial success of valency control in hydrogenated amorphous silicon (a-Si:H) led by Spear and LeComber [1], this material has been extensively studied for thin film device applications such as low cost solar cells, thin film transistors and photoreceptors in xerography. In general, a-Si:H is prepared by glow-discharge decomposition of SiH$_4$ (or Si$_2$H$_6$), and, as is well known, the film properties vary depending strongly on the conditions of SiH$_4$ (or Si$_2$H$_6$) glow-discharge plasma. For the past seven years a variety of plasma diagnostic techniques have been applied to the SiH$_4$ plasma; optical emission spectroscopy (OES) [2], ion mass spectrometry (MS) [2], infrared emission and absorption spectroscopy [3], laser induced fluorescence (LIF) [4] and coherent anti-Stokes Raman spectroscopy (CARS) [5].

This paper describes our recent OES, MS and CARS measurements on Si$_2$H$_6$ as well as SiH$_4$ glow-discharge plasmas, and discusses the dominant species involved in each plasma in relation to the deposition kinetics of a-Si:H.

EXPERIMENTAL

OES, MS and CARS were employed in the present work as diagnostic tools for SiH$_4$ and Si$_2$H$_6$ plasmas: OES (optical emission spectroscopy) identifies emissive species involved in the plasma such as H*, H$_2$*, Si* and SiH* and determines their densities (excited states); MS (mass spectrometry) gives the information on ionic species (H$^+$, H$_2^+$, Si$_x$H$_y$....); and CARS (coherent anti-Stokes Raman spectroscopy) detects non-emissive species (ground states) such as SiH_4, Si_2H_6, SiH_2 and H_2 with a highly-resolved spatial distribution of their densities.

A capacitively-coupled rf (13.56 MHz) glow-discharge reactor was used to produce SiH_4 and Si_2H_6 plasmas, which underwent OES, MS and CARS measurements. Details of the experimental setup were described earlier [2][5]. Except where noted, 100-% SiH_4 (or Si_2H_6) was fed at a flow rate of 5-30SCCM with a gas pressure of 50-200 mTorr, and rf power supplied was controlled in the range of 0.01-1 W/cm^2. All of the data taken from the plasma were discussed in connection with the deposition rates of resulting a-Si:H films.

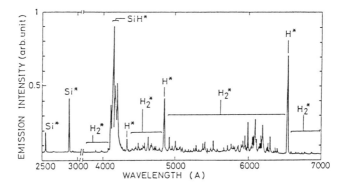

Figure 1 − Typical OES data of SiH₄ glow-discharge plasma.

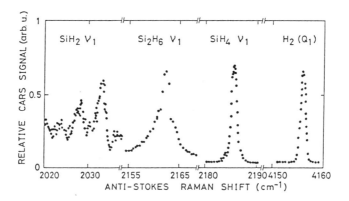

Figure 2 − Observed CARS signals from SiH₄ and Si₂H₆ plasma.

RESULTS AND DISCUSSION

Figure 1 shows a typical OES data of a pure SiH_4 glow-discharge plasma traced in the range of 250 nm - 700 nm. The emission lines at 2881A (UV 43 transition: Si*), 4127A ($A^2\Delta - X^2\pi$: SiH*), 6021A ($3p^3\pi-2s^3\Sigma$: H_2*) and 6563A ($3^2D-2^2p^0$: H*) were mainly used in this work [2]. Figure 2 shows CARS spectra identified with SiH_2 ν_1 (2034 cm^{-1}), Si_2H_6 ν_1 (2162 cm^{-1}), SiH_4 ν_1 (2187 cm^{-1}) and H_2 (4157 cm^{-1}) bands, respectively. The signal intensity is quadratically proportional to the magnitude of the third order electric susceptibility, which is in turn linear to the number of molecules (ground state) in resonance [5].

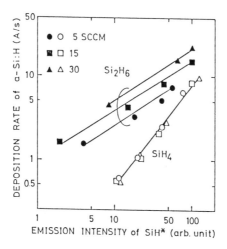

Figure 3 — Deposition rate of a-Si:H from SiH_4 and Si_2H_6 plasma as functions of SiH* emission intensity.

Figure 3 shows the deposition rate of a-Si:H film from Si_2H_6 and SiH_4 glow discharge plotted against the emission line intensity from SiH* radicals. As is clear in the figure, the deposition rate of a-Si:H from SiH_4 plasma is proportional to the line intensity of SiH* over a wide range of parameters, i. e.,

$$\text{Dep. rate } (SiH_4) \quad \propto \quad [SiH^*] \qquad\qquad (1)$$

seems to hold in deposition kinetics from the SiH_4 plasma [6]. However, according to the luminosity measurement of a 4127-A light (SiH*) from SiH_4 plasma (0.01 W/cm^2, 5 SCCM, 50 mTorr) the density of excited SiH [SiH*] has been estimated as

$3 \times 10^6 / cm^3$, which is by three or four orders of magnitude lower than the density of species required for explaining the observed deposition rate. It implies that SiH* is not a direct precursor for a-Si:H deposition but other species strongly correlated with SiH* are responsible.

In contrast to SiH₄ plasma, a relation is not so simple between the SiH* intensity from Si₂H₆ plasma and the deposition rate of resultant a-Si:H film, although the behavior could be fitted to a relation

$$\text{Dep. rate (Si}_2\text{H}_6) \propto [\text{SiH}^*]^{1/2}, \tag{2}$$

if the flow rate of Si₂H₆ is kept constant (see Fig.3).

In order to elucidate the relations represented by Eqs.(1) and (2), OES and CARS signals from relevant species were measured as functions of time after turning on the rf power supplied to the SiH₄ as well as Si₂H₆ gas in the closed reactor. The results are shown in Figs.4 and 5.

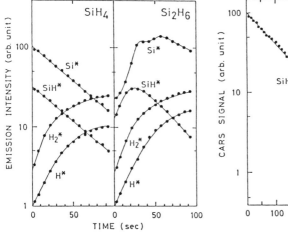

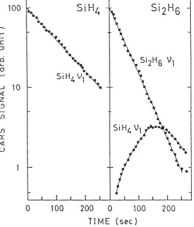

Figure 4 − Time-dependent emission intensity from each species in closed SiH₄ and Si₂H₆ plasma (0.25 W/cm², 100 mTorr).

Figure 5 − Time-dependent CARS signals in closed SiH₄ and Si₂H₆ plasma (0.05 W/cm², 120 mTorr).

As for OES data in Fig.4, behavior of SiH* and Si* are quite different between SiH₄ and Si₂H₆ glow discharge, while those of H* and H₂* are similar in both plasmas. Namely, [SiH*] as well as [Si*] decays monotonously in the closed SiH₄

plasma,   but,   in Si_2H_6 plasma, both line intensities initially
increase and afterwards decrease (with time). This tendency is
strongly correlated with the time-dependent behavior of the
signal intensity of CARS SiH_4 ν_1 Q band in SiH_4 and Si_2H_6
plasmas.   As is evident in Fig.5, the number of SiH_4 species
decrease at a constant loss rate in SiH_4 plasma simply due to
the electron-impact dissociation,   whereas in Si_2H_6 plasma SiH_4
species are produced via electron-impact dissociation of Si_2H_6
molecule or subsequent secondary chemical processes.

The above arguments based on Figs.4 and 5 strongly
suggest that SiH* as well as Si* emission originates from a
direct dissociation process of super-excited SiH_4 molecule not
only in SiH_4 but also in Si_2H_6 glow discharge plasmas.   In
other words, SiH* emission in SiH_4 plasma should be ascribed to
one-electron-impact excitation process,   whereas SiH* emission
requires at least two-electron-impact excitation process in
Si_2H_6 plasma because no SiH_4 molecules are involved in starting
Si_2H_6.   This tentative discussion is compatible with the
previous works done by several groups independently as for the
emission process of SiH* in SiH_4 plasma [2][4].

All of these results involving the relations of Eqs.(1)
and (2) give us important information on the generation
kinetics of dominant precursors for a-Si:H film deposition in
SiH_4 as well as Si_2H_6 plasmas: (1) The main film precursor in
SiH_4 plasma is produced through primary one-electron-impact
dissociation of SiH_4 molecule,   and its concentration should be
proportional to [SiH*].   (2) The dominant film precursor in
Si_2H_6 plasma seems to be generated also by primary electron-
impact dissociation process of Si_2H_6 molecule,   because the
deposition rate is proportional to the square root of SiH*
intensity whose emission might be initiated by two-electron
impact excitation.   It also implies that dominant precursors
are different between SiH_4 and Si_2H_6 glow discharge.

The available data are insufficient to provide a distinct
assignment of the film precursors in both plasmas;   however
several positive discussions can be made on the basis of other
independent experiments (below).

Recently,   Schmitt has reported cross sections for
electron impact on SiH_4 as a function of an electron energy and
determined the branching ratio of SiH_4 into various species
induced by 70-eV electron impact [4].   Although ionization and
neutral dissociation are of comparable importance in this high
electron energy range, the amount of ionic species is much less
than that of neutral radicals in a realistic deposition
condition because an average kinetic energy of electron lies in
the range between 3 and 10eV,   which has been confirmed by our
MS measurements [2].   The most striking indication of the work
done by Schmitt et al. is that SiH_3 and/or SiH_2 might dominate
in SiH_4 plasma and could be candidates for main precursors of
a-Si:H deposition,   while,   according to LIF (laser-induced

fluorescence) measurement, ground state SiH is a minor component among the created radicals in the primary processes.

On the other hand, according to our CARS measurement, SiH$_2$ radicals have been observed as SiH$_2$ ν_1 Q band (shown in Fig.2) only when starting SiH$_4$ gas is diluted by H$_2$ (H$_2$/SiH$_4$=9/1) [5]. No signals were detected from 100-% SiH$_4$ glow discharge. Namely, the reaction lifetime of SiH$_2$ in 100-% SiH$_4$ plasma is quite short, suggesting that SiH$_3$ is most probable film precursor in 100-% SiH$_4$ glow-discharge plasma, because SiH$_3$, different from SiH$_2$, does not react with silane.

In order to determine the reaction lifetime (τ) of each dominant precursor in each plasma, a specially-designed triode reactor was employed [6]. The values of τ of main precursors were estimated as 1.6×10^{-3} sec for SiH$_4$ plasma and 1.5×10^{-4} sec for Si$_2$H$_6$ plasma. The shorter reaction-time precursor in Si$_2$H$_6$ is considered to be Si$_2$H$_y$ neutral radicals, since the MS measurement have shown that Si$_2$H$_y^+$ species are dominant in the plasma while SiH$_x^+$ in SiH$_4$ plasma.

In summary, neutral species (emissive as well as non-emissive) and ionic species involved in SiH$_4$ and Si$_2$H$_6$ glow discharge plasma have been detected using OES, MS and CARS techniques.The main precursor for a-Si:H deposition is produced via a primary one-electron-impact dissociation of the starting gas molecule (SiH$_4$ as well as Si$_2$H$_6$). SiH$_3$ and Si$_2$H$_y$ have been suggested as most probable precursors for a-Si:H deposition from SiH$_4$ and Si$_2$H$_6$ plasma, respectively.

REFERENCES

1. Spear, W.E.; LeComber, P.G. Substitutional doping of amorphous silicon. Solid State Commun. 1975, 17, 1193-1196.
2. Matsuda, A.; Tanaka, K. Plasma spectroscopy – glow discharge deposition of hydrogenated amorphous silicon. Thin Solid Films 1982, 92, 171-187.
3. Knights, J.C.; Schmitt, J.P.M.; Perrin, J.; Guelachvilli, G. High resolution absorption and emission spectroscopy of a silane plasma in the 1800-2300cm^{-1} range. J. Chem. Phys. 1982, 76, 3414-3421.
4. Schmitt, J.P.M. Fundamental mechanisms in silane plasma decompositions and amorphous silicon deposition. J. Non-cryst. Solids 1983, 59/60, 649-658.
5. Hata, N.; Matsuda, A.; Tanaka, K. Neutral radical detection in silane glow-discharge plasma using coherent anti-Stokes Raman spectroscopy. J. Non-cryst. Solids 1983, 59/60, 667-670.
6. Matsuda, A.; Kaga, T.; Tanaka, H.; Tanaka, K. Lifetime of dominant radicals for the deposition of a-Si:H from SiH$_4$ and Si$_2$H$_6$ glow discharges. J. Non-cryst. Solids 1983, 56/60, 687-690.

CHAPTER 25

THE CONVERSION OF SOME POLYORGANOSILICON PRECURSORS TO CERAMICS

Ronald H. Baney, Gary T. Burns, John P. Cannady, and Terrance K. Hilty
Dow Corning Corporation, Midland, Michigan, USA

INTRODUCTION

Recently there has been enhanced interest in silicon carbide and silicon nitride as high temperature structural ceramics, particularly for heat engine applications [1]. Commercial utilization of these types of ceramic has been hindered by the difficulty of forming complex shapes and their inherent brittle nature. German [2] and Japanese researchers [3] have developed organosilicon polymers which are capable of being spun into fibers and then pyrolyzed to afford silicon carbonitride and silicon carbide ceramic fibers. These were prepared via processes similar to the procedures employed for the preparation of carbon fibers [4]. These pioneering works have led to a flurry of activity by other researchers [5].

The conversion of organosilicon polymers to ceramic fibers is a complex process involving formation of the polymer, spinning, cure and pyrolysis. A large amount of literature has dealt with the chemistry of preceramic polymer preparations. Much less has dealt with the important pyrolysis step to convert the organosilicon polymer to inorganic ceramics. · Methylpolysilane systems were prepared from the catalyzed SiSi/SiCl redistribution of methylchlorodisilanes and methylpolydisilylazanes from the SiN/SiCl redistribution of

methylchlorodisilanes and hexamethyldisilazane.
The pyrolysis of these systems is presented below
to illustrate the chemistry of the conversion of a
polymer into a ceramic.

SYNTHESIS OF PRECERAMIC POLYMERS

Methylchloropolysilane (MPS-Cl) was prepared
according to the method of Baney, et al., [6-7]. A
mixture of freshly distilled methylchlorodisilanes
consisting of 11% $(Me_2ClSi)_2$; 30% $Me_2ClSiSiCl_2Me$
and 58% $(MeCl_2Si)_2$ was slowly heated with 1% Bu_4PCl
to 250 C while $MeSiCl_3$ and Me_2SiCl_2 distilled from
the reaction giving a polymeric residue with a ^1H
NMR spectra and an IR spectra identical to that
described earlier [7] which was shown to have an
approximate formula of $(Me_2Si)_3(MeSi)_{17}Cl_5$.

The chlorine was substituted for a methoxy
group by reacting the chlorine containing polymer
described above with excess methylorthoformate in
toluene [8]. The resulting polymer (MPS-OMe) had a
molecular weight of 1400 as determined by GPC. The
^1H NMR in $CDCl_3$ gave a broad peak at $\delta = 3.54$ ppm
assigned to the $SiOCH_3$. The $SiOCH_3/SiCH_3$ ratio
from the NMR analysis was 1/42.

The chlorine was also substituted with phenyl
by reacting a toluene solution of the chlorine
containing methylpolysilane with an excess of
phenylmagnesium bromide similar to literature
procedures [9]. After hydrolysis with aqueous
NH_4Cl, the resulting polymer(MPS-Ph) had the
following ^1H NMR and IR. NMR ($CDCL_3$) δ 0.49
(center of broad peaks), 0.86 (s), 7.06 - 7.69 (m,
Si-Ph). The SiMe:SiPh ratio was 31.8:2.6). IR
(KBr): 3056 (w), 2959 (m), 2894 (m), 1403 (m), 1244
(m), 1016 (s), 773 (s), 698 (s).

In order to prevent the polymers from melting
and losing shape after spinning some crosslinking
or "cure" reaction is required. One such reaction
is the hydrolysis of the methoxy group in the
methoxy-containing methylpolysilane. To a
tetrahydrofuran solution of the polymer was added
as excess of H_2O containing HCl. The reaction
mixture was stirred until gellation occurred. The
resulting precipitate (MPS-O) was collected as
dried under vacuum. IR (KBr) of the resulting
gelled polymer was 3402 cm^{-1} (s), 2959 (m), 2894

(m), 2156 (m), 2073 (m), 1409 (m), 1260 (s),769 (vs).

Methylpolydisilazane was prepared by reacting mixed disilanes similar to those in the preparation of methylchloropolysilane described above with excess hexamethyldisilazane. Mixed disilanes with an empirical formula ($Me_{2.6}Cl_{3.4}Si_{2.0}$) were purified by distillation under an argon atmosphere. This methylchlorodisilane mixture was then combined with an excess of hexamethyldisilazane. The cloudy mixture containing small amounts of ammonium chloride (resulting from inadvertent hydrolysis reactions) was slowly heated to 190 C. The ammonium chloride, trimethylchlorosilane and excess hexamethyldisilazane was distilled from the reaction mass. A clear, colorless, brittle preceramic polymer was obtained [10]. The resulting polymer, methylpolydisilylazane MPDZ, had an approximate formula of
$[(Me)_{2.6}(Si_2)_{1.0}NH_{1.5}(NHSiMe_3)_{0.4}]_x$ [11]. The polymer was also shown to contain about 4% Cl as unreacted $\equiv SiCl$ moieties or NH_4Cl.

PYROLYSIS OF THE METHYLPOLYSILANE SYSTEMS

The char yield and oxygen content of the polymers fired to 1200 C in an argon atomsphere at 10 degrees per minute and held for an half hour at 1200 C were: MPS-Cl (44% C.Y., 2.5% O); MPS-OMe (63% C.Y., 9.5% O); MPS-Ph (61% C.Y., 5.7% O); MPS-O (82% C.Y., 9.2% O).

The higher char yield for the methoxy and phenyl derivatives over the chloro derivative is probably the result of inadvertent hydrolysis during work up. This would result in higher crosslinking, which would reduce the vaporization of volatile oligomers before the pyrolysis reaction occurs. The higher oxygen content for the high char yield systems are consistent with this hypothesis.

Infrared studies were carried out and the pyrolysis reactions of the four polymers with the aid of a Nicolet model 5DX FTIR and a Pyro-Chem Foxboro pyrolysis gas cell model 40 was used for pyrolysis gas studies. When the four polymers were heated to 440 C at a heating rate of 3 C per minute, a very intense peak developed in the infrared spectra (KBr pellet) at 2090 cm^{-1}. A weaker peak at about 1350 cm^{-1} also developed for all of the polymers. The first peak has been assigned to the SiH asymmetric stretching mode and the weaker peak to the $SiCH_2Si$ deformation mode. The presence of these peaks during intermediate stages of pyrolysis probably arise from a Kumada type rearrangement [12].

$$CH_3SiSi \xrightarrow{\quad \Delta \quad} HSiCH_2Si$$

Small amounts of SiH species were reported earlier in the pyrolysis gases of PCP-250-Cl and tentatively attributed to the Kumada type rearrangement [12]. Nujol mulls were prepared from the ceramic char after pyrolysis to 1200 C. All four polymer chars had similar IR spectra with a broad peak from 1000 cm^{-1} to 400 cm^{-1}. This in contrast to distinct peaks 1080 cm^{-1} and 460 cm^{-1} assigned to SiC observed by Hasegawa and Okamura [13] in the "Yajima" polycarbosilane.

Pyrolysis gases were examined in a pyrolysis gas IR cell for the four polymers. At about 400 C methane was observed for all four polymers. Absorptions at 3017 cm^{-1} and 2900 cm^{-1} (HCl) were also observed for the MPS-Cl sample. No benzene was observed in the pyrolysis gases of the MPS-Ph even to pyrolysis temperatures of 1000 C. Small amounts of CO (2164 cm^{-1} and 2123 cm^{-1}) were also observed for all of the samples. A peak at 1048 cm^{-1} observed in the MPS-Cl was assigned to SiO-Si compounds formed inadvertent hydrolysis during sample preparation (largely in an argon controlled atmosphere box).

PYROLYSIS OF METHYLPOLYDISILYLAZANE

IR spectra (KBr pellet) of methylpoly-disilylazane were taken after heating the polymer in 100 C increments to 1300 C in argon for four hours. At about 300 C bonds for SiH ~2100 cm^{-1} and 1350 cm^{-1} $SiCH_2Si$ began to appear as in the methylpolysilane systems described above. A band at 3362 cm^{-1} assigned to NH asymmetric stretch in the 300-500 C region decreased suggesting formation of Si_3N moieties. Between 500 and 800 C, bonds associated with SiH and CH disappear. Previously reported studies [14] on this system had indicated methane as the primary gas formed in this temperature region. From 800 C to 1200 C a broad featureless band between 1400 cm^{-1} and 600 cm^{-1} remains. A TGA study reported earlier on the same pyrolysis [14] shows breaks in the same three regions as distinct changes occur in the infrared of the char.

A char yield of 53% was observed for methylpolydisilylazane polymers polymerized to 250 C. The density of the char was 2.2 Mg/m^3 and the char was analyzed to have an empirical formula of $Si_{1.0}C_{0.8}N_{0.7}O_{0.05}$.

REFERENCES

1. J. J. Burke, A. E. Gorum and R. N. Katz, "Ceramics for High Performance Applications", No. 2 Proc. 2nd Army Materials Tech. Conf. Brook Hill Publishing Co., Chestnut Hill, Massachusetts, USA (1974).

2. W. Verbeek, "Materials Derived from Homogeneous Mixtures of Silicon Carbide and Silicon Nitride and Methods for Their Production," U.S. Patent No. 3,853,567, 1974. W. Verbeek and G. Winter, "Silicon Carbide Shaped Article and Process for the Manufacture Thereof," Ger. Offen. 2,286,078, Bayer A. G., March 2q, 1974; G. Winter, W. Verbeek, and M. Mansmann, "Production of Shaped Articles of Silicon Carbide and Silicon Nitride," Ger. Offen. 2,243,527, May 16, 1974 (U.S. Patent No. 3,892,583).

3. S. Yajima, J. Hayashi, M. Omori, and K. Okamura "Development of a Silicon Carbide Fiber with High Tensile Strength," Nature 261, 683-685 (1976); S. Yajima, J. Hayashi, and M. Omori, "Silicon Carbide Fibers having a High Strength and a Method for Producing Said Fibers," U.S. Patent No. 4,100,233, July 11, 1978; S. Yajima, J. Hayashi, and K. Okamurs, "Pyrolysis of a Poly-(borodiphenylsiloxane)," Nature 266, 521-522 (1977).

4. A. Shindo, "Osaka Hoyya Gijuisu Shikensho Hokoku," Rept. Govt. Ind. Res. Inst. Osaka, No. 317 (1961).

5. Pre-Ceramic Polymers - American Ceramic Society Bulletin, 62 (8) 889-923 (1983).

6. Ronald H. Baney and John H. Gaul, Jr., "Method of Preparing Silicon Carbide", U.S. Patent 4,310,651, Jan. 12 (1982).

7. Ronald H. Baney, John H. Gaul, Jr., and Terrence K. Hilty, "Methylchloropoly-silanes and Deriviatives Prepared from the Redistribution of Methylchlorodisilanes", Organometallics, 2,859, (1982).

8. Ronald H. Baney and John H. Gaul, Jr., "High Yield Silicon Carbide Preceramic Polymers", U.S. Patent 4,298,558, Nov. 3, (1981).

9. Ronald H. Baney and John H. Gaul, Jr., "High Yield Silicon Carbide From Alkylated or Arylated Preceramic Polymers, U.S. Patent 4,298,559, Nov. 3, (1981).

10. J. H. Gaul, Jr., "Process For The Preparation of Poly(disilyl)silazane Polymers And The Polymers Therefrom", U.S. Patent 4,340,619, July 20, (1982).

11. L. L. Hench and D. R. Ulrich, "Ultrastructure Processing of Ceramic Glasses and Composites", Chap. 20. R. Baney, "Some Organometallic Routes to Ceramics", John Wiley and Sons, New York, NY, (1984).

12. K. Shinu and M. Kumada, "Thermal Rearrangement of Hexamethyldisilane to Trimethyl(dimethyl-silymethyl)silane," J. Org. Chem. 23, 139 (1958). H. Saduraj, A. Hasomi, and M. Kumada, "Thermolysis of Hexamethyldisilane," Chem. Commun. 16, 930-2 (1968).

13. Y. Hasegawa and K. Okamura, "Synthesis of Continuous Silicon Carbide Fibers," part 3, J. Mat. Sci. 18 3633-3648 (1983).

14. Ronald H. Baney, Polymer Preprints, 25 (1) 2-3 Apl. (1984).

CHAPTER 26

SYNTHESES AND CATALYTIC ACTIONS OF ORGANOSILICON POLYMER-SUPPORTED METAL COMPLEXES

Mei-Yu Huang, Chang-Yu Ren, Yan-Zhu Zhou, Li Zhao, Xian-Peng Cao, Diao-Ling Dong, Yun Liu and Ying-Yan Jiang* Institute of Chemistry, Academia Sinica, Zhongguancun, Peking 100080, China

SYNOPSIS

Several organosilicon polymer-supported palladium, platinum and rhodium complexes have been prepared and used as hydrogenation catalysts for olefines, aldehydes, ketones, nitro-compounds and aromatics; hydrosilylation catalysts for olefines and acetylene; oxidation catalysts for methanol, ethanol and isopropanol; and hydroformylation catalyst for heptene at mild conditions respectively.

INTRODUCTION

Many highly active and selective catalysts can be derived from organometallic complexes. These homogeneous complex catalysts have several advantages over conventional heterogeneous catalysts in that the active centres are all accessible to reagents and their properties can be controlled in a systematic manner by variations in the ligand groups attached to the transition metal. However, these homogeneous catalysts are often unstable, corrosive and difficult to recover from the reaction products. In order to overcome these disadvantages, intensive research on polymer-supported metal complex catalysts has been carried out during the past decade. The polymer supports of polymer-supported metal complexes can be classified into two types, i.e., organic polymers containing C-C bonds as the main chain and organosilicon polymers containing Si-O bonds as the main chain. The former has been researched in more than the latter.

Recently, we prepared some organosilicon polymer-supported metal complexes and used them as hetrogeneous hydrogenation, hydrogenolysis, hydrosilylation, oxidation and hydroformylation catalysts respectively, and have found that they had more advantages than the organic polymer-supported metal complexes.

HYDROGENATION AND HYDROGENOLYSIS CATALYSTS

Several organic polymer-supported metal complexes [1-4] have been shown to be more active and selective than the corresponding organometallic complexes for hydrogenation of olefines at room temperature and under atmospheric pressure. However, until several years ago, no organosilicon polymer-supported metal complex had been shown to have such high catalytic activity. For example, silica-poly-β-cyanoethyl-siloxane-Pd complex [5] was shown to be not as active as Pd itself for the hydrogenation of olefins. Afterwards, we improved the preparation method of organosilicon polymer-supported metal complexes to increase catalytic activity. For example, Pd complexes of silica-poly-γ-cyanopropylsiloxane (abbreviated as Si-CN) [6] and silica-poly-γ-aminopropylsiloxane(Si-NH₂) [7] ,and Pt complex of silica-poly-γ-diphenylphosphinopropylsiloxane (Si-P) [8] have been prepared and found to be active hydrogenation catalysts for olefines, and Si-NH₂-Pd and Si-P-Pt were more active than any other organic polymer-supported complexes [1-4] . Si-P-Pt was found very stable and could be reused thirty times without any appreciable loss in the catalytic activity.

$$\text{SiO}_2\!\!-\!\!-\text{O}-\!\!\underset{\overset{|}{\text{O}}}{\overset{|}{\text{Si}}}\!-(\text{CH}_2)_3\,\text{CN}$$
(Si-CN)

$$\text{SiO}_2\!\!-\!\!-\text{O}-\!\!\underset{\overset{|}{\text{O}}}{\overset{|}{\text{Si}}}\!-(\text{CH}_2)_3\,\text{NH}_2$$
(Si-NH₂)

$$\text{SiO}_2\!\!-\!\!-\text{O}-\!\!\underset{\overset{|}{\text{O}}}{\overset{|}{\text{Si}}}\!-(\text{CH}_2)_3\,\text{PPh}_2$$
(Si-P)

Such high stability has not been found for organic polymer-supported metal complexes reported before.

Some organometallic complexes and only one organic polymer-supported metal complex, Rh complex of polystyrene with anthranilic acid groups [9] have been shown to catalyze aldehydes, ketones, nitro-compounds or aromatics, but almost all required high temperature and high pressure as reaction conditions. We have found Si-CN-Pd and Si-NH₂-Pd to be the effective hydrogenation catalysts for various kinds of aldehydes and ketones at room temperature and under atmospheric pressure. The products are alcohols. For example, benzaldehyde and acetophenon can be converted to benzylalcohol and d-phenylethanol in 100% yields, respectively[10] . We have shown that two kinds of organic polymer-supported metal complexes, Pd complexes of silica-polyacrylonitrile[11] and poly-vinylpyrrolidone[12] could catalyze the hydrogenation of various aromatic nitro-compounds at room temperature and under atmospheric pressure. For example, nitrobenzene can be reduced to aniline in 100% yield. However, they could not catalyze the hydrogenation of aliphatic nitro-compounds under the same conditions. Recently, we have found that Si-CN-Pd, Si-NH2-Pd, silica-poly-γ-carboxypropylsiloxane-Pd complex (Si-COOH-Pd) and silica-poly-γ-diacetamidopropylsilo-xane-Pd complex (Si-NCOCH3-Pd) could catalyze not only aromatic nitro-compounds but also alkyl nitro-compounds such as nitromethane and nitroethane.

$$\text{SiO}_2 \rangle\!\!- -O-\overset{|}{\underset{\overset{\|}{O}}{Si}}-(CH_2)_3\,COOH$$

(Si-COOH)

$$\text{SiO}_2 \rangle\!\!- -O-\overset{|}{\underset{\overset{\|}{O}}{Si}}-(CH_2)_3\,N(COCH_3)_2$$

(Si-NCOCH3)

Moreover, when Si-COOH-Pd or Si-NCOCH3-Pd was used as hydrogenation catalyst for benzaldehyde and acetophenone, the products were not only benzylalcohol and d-phenylethanol but also toluene and ethyl benzene respectively. In these cases, hydrogenolysis also occured in addition to hydrogenation. We have prepared some organic polymer-supported metal complexes such as Rh complexes of polyamide[13] , polyvinyl chloride with dimethyl-amino[14] or morpholino[15] groups, and have demostrated that they could catalyze benzene hydrogenation at room temperature and under atmospheric pressure. Unfortunately, these catalysts were not so stable during hydrogenation. Afterward, Si-P-Pt was found to be a very active, selective and stable hydrogenation and hydrogenolysis catalyst for vario-us kinds of aromatics in the presence of a small amount of hydrochloric acid(as a promotor under the same conditions.[16,17]   For example, the hydrogena-tion of benzene, toluene and phenol gave cyclohexane , methylcyclohexane and cyclohexanol in 100% yields respectively.   In the hydrogenation of chloroben-zene, bromobenzene and hydroquinone, the hydrogeno-lysis also occured, so the products were cyclohexane, cyclohexane and cyclohexanol respectively.   It is noteworthy that in the hydrogenation of benzaldehyde and acetophenone, not only the benzene rings but the aldehydo and keto groups were also hydrogenated   and cyclohexyl carbinol and d-cyclohexyl ethanol were produced, respectively. The yields were all 100%. Recently, we have prepared a Pt complex of silica-polyphenylsilazane (PhSi-N-Pt) and have demonstrated that it could catalyze the hydrogenation and hydrogenolysis of various aromatics in the absence of any promoter.(at room temperature and under atmospheric pressure).

SiO_2— —$\overset{C_6H_5}{\underset{HN}{Si}}$——N-

(PhSi-N)

HYDROSILYLATION CATALYSTS

Silica-poly-γ-aminopropylsiloxane-Pt complex (Si-NH2-Pt)[16] and silica-poly-γ-mercaptopropylsiloxane-Pt complex (Si-SH-Pt)[19,20] have been prepared and found that they could catalyze the addition of triethoxysilane to 1-hexene at 80°C or room temperature to give n-hexyltriethoxysilane in over 95% yield respectively.

SiO_2— -O-$\overset{|}{\underset{O}{Si}}$-(CH2)3SH

(Si-SH)

The latter (turnover numbers, 10,000) was more stable than the former (turnover number, 2,000). during the reaction. The addition of triethoxysilane to acetylene catalyzed by Si-SH-Pt at 80°C or room temperature gave vinyltriethoxysilane and bis(triethoxysilyl)ethane in 50-60 and 20-50% yields respectively.

OXIDATION CATALYSTS

Formaldehyde is manufactured by oxidation of methanol with MoO_3-Fe_2O_3 , Cu-Zn or Ag at high temperature(300-750°C). If phenanthroline-Cu complex was used as catalyst, methanol could be oxidized to formaldehyde at 30-45°C [21], However, in this process, a large amount of formic acid was produced at the same time, and the reaction would stop spontaneously if formic acid was not neutralized by alkali in time. We have found that Si-P-Pt or PhSi-N-Pt could catalyze the oxidation of methnol to give formaldehyde quantitatively at room temperature and under an atmospheric oxygen pressure. No by-product was detected.

Acetone is manufactured by oxidation of isopropanol in the presence of copper at high temperature (400-500°C). If a sulfonated polystyrene-Co complex was used as catalyst, isopropanol could be oxidized to acetone at 50°C . However, the yield

of acetone was only 5-6%[22].   We have found that
Si-P-Pt and Pt complex of silica-polysilazane
(Si-N-Pt) could catalyze the oxidation of isopropanol to give acetone in 50 and
90% yields respectively at room
temperature and under an atmospheric oxygen pressure.
            The manufacture of acetal-
dehyde or acetic acid from
oxidation of ethanol catalyzed by Ag also requires
high temperature (500°C).   Although [Cu(4,4'-Me$_2$-
-bpy)$_2$] Cl or [Cu(bpy)$_2$]Cl can catalyze the oxidation
of ethanol to acetaldehyde at 15°C.   they are
very unstable, and loose their catalyticabilities
after 10 hours[23].   We have found that Si-N-Pt could
catalyze the oxidation of ethanol to acetaldehyde in
90% yield, and Si-P-Pt could catalyze the oxidation
of ethanol to acetic acid in 100% yield.

(Si-N)

HYDROFORMYLATION CATALYSTS

[24] Phosphinated organic polymer-supported Rh complex
has been shown to catalyze the hydroformylation
of olefines.   However, the phosphine group is
oxidized in the presence of oxygen to cause the
eduction of rhodium.   Therefore, a nonphosphine
anchored polymer-supported Rh complex, such as Rh
complex of polystyrene with iminodiacetic acid
groups has been used as hydroformylation catalyst of
hexene-1.   However, in this case, acetal was
produced as by-product, the yield of aldehyde was
only about 60%, and the eduction of rhodium was also
observed after the reaction[25].
    Recently, we have found that Si-N-Rh, MeSi-N-Rh
and PhSi-N-Rh could catalyze the hydroformylation
of heptene-1 at 110°C and
and under 30 atm CO and H$_2$
for 10 hours to give
octanal in 96, 95 and 32%
yields, and n/i ratio in
0.51, 0.71 and 2.2 respec-

(MeSi-N-Rh)

tively.   An extraodinary property of Si-N-Rh or
Me-Si-N-Rh is their high stability in hydroformylation process.   Si-N-Rh and MeSi-N-Rh could be
reused over five and Six times, and the total turnover numbers were over 120,000 and 130,000 respectively.

REFERENCES

(1) R.H.Grubbs,L.C.Kroll,J.Am.Chem.Soc.,93,3062(1971)

(2) R.H.Grubbs,L.C.Kroll,E.M.Sweet, J.Macromol.
    Sci-Chem., A. 7, 1047(1973).
(3) N.Nakamura,H.Hirai, Chemistry Letters, 645,809
    (1974); 823(1975); 165(1976).
(4) K.Kaneda et al.,Chemistry Letters, 1005(1975).
(5) V.A.Semikolenov,V.A.Likholobov,Yu.I.Ermakov,
    Kinet.Katal., 18, 1294(1977).
(6) Y.Z.Zhou,Y.Y.Jiang,Polymer Preprints, Japan,
    28, 2018(1975).
(7) Y.Z.Zhou,Y.Y.Jiang,Polymer Comm.,China,97(1979).
(8) Y.Z.Zhou,Y.Y.Jiang,J.Catal.,China, 2,233(1981).
(9) N.J.Holy,J.Org.Chem., 44,239(1979).
(10)Y.Z.Zhou,Y.Y.Jiang,J.Organometal.Chem.,251,31
    (1983).
(11)Y.J.Li,Y.Y.Jiang,J.Catal.,China, 2, 42(1981).
(12) Y.J.Li,Y.Y.Jiang,J.Mol.Catal.,19, 277(1983).
(13)Z.H.Chen,Y.Y.Jiang,Chemical Reagents,China,
    165(1980).
(14)Z.H.Chen,Y.Y.Jiang, J.Catal.,China,2,149(1981).
(15)Y.Z.Zhou,Z.H.Chen,Y.Y.Jiang,Preprints of Japan-
    -China Bilateral Symposium on Polymer Science and
    Technology, Tokyo-Kyoto,Japan, 299(1981).
(16)Y.Z.Zhou,Y.Y.Jiang,J.Mol.Catal.,19,283(1983).
(17)Y.Y.Jiang, The New Chemical Materials,China,
    12, (1), 1(1984).
(18)L.Z.Wang,Y.Y.Jiang,J.Catal.,China,2,236(1981).
(19)L.Z.Wang,Y.Y.Jiang,J.Organometal.Chem.,251,
    39(1983).
(20)D.X.Wan,L.Z.Wang,Y.Y.Jiang,Polymer Comm.,
    China, 235(1982).
(21)W.Brackman,C.J.Gaasbeek,Recl.Trav.Chim.,
    85,242(1966).
(22)J.Maternova et al., J.Polymer Sci. Polymer
    Symposium, no.68, 239(1980).
(23)M.Munakata et al., J.Chem.Soc.Chem.Comm.,
    219(1980).
(24)C.U.Pittman,Jr.,R.M.Hanes,J.Am.Chem.Soc.,
    98, 5402(1976).
(25)H.Hirai et al., J.Chem.Soc.,Japan, 122,316
    (1982).

CHAPTER 27

CURRENT TRENDS OF SILICONE INDUSTRY IN JAPAN

Tadashi Wada Shin-Etsu Chemical Co., Ltd., Tokyo, Japan

INTRODUCTION

Starting with the study conducted by F.S. Kipping early this century, development of silicone has progressed since the 1930's as a heat-resistant insulating material. The Direct Method invented by E.G. Rochow established the means for its production on a commercial basis.

In the beginning, silicone found its only application as a polymer material in a limited number of sectors such as munitions. As technology has advanced, its original characteristics have become appreciated because they are somewhat different from other common organic polymers. New applications of silicone are being found regularly, many closely related to meeting the high performance and versatile needs of the frontier industries of the times. New products and applications of silicone are being actively developed.

This paper places greater emphasis upon details of the technological development of the Japanese silicone industry as related to the development of other Japanese industrial technology than upon the current trends of the Japanese silicone industry as suggested by the title.

DETAILS OF THE DEVELOPMENT OF THE JAPANESE SILICONE INDUSTRY

Development of the Japanese silicone industry will be described in detail hereunder in terms of the trends of main products and applications of silicone based on chronology.

The 1950s

This period was marked by Japan´s postwar rehabilitation as well as the creation of the silicone industry. The silicone industry in Japan began production in 1953. Demands for silicone mainly came from the heavy electric and textile industries, finding applications as insulating varnishes for motors and transformers and as water repellents for textiles.

The 1960s

Japan had remarkable economic growth in this period, mainly in the heavy chemical industry. Regarding silicone, new technology and new products were successively developed, resulting in expanded demand. The leading industries of those days included the automobile and color TV industries. Taking these two lines of products as an example, how silicone contributed to improvements in their quality and productivity will be described.

In the automobile industry, silicone rubber was used as an oil seal material for transmission shafts and crankshafts, etc. Silicone rubber has superior dynamic properties with excellent high sealing performance for devices requiring high speed rotation. In addition, silicone is characterized by very little variation in physical properties over a wide temperature range, and thus is highly heat-resistant. At the same time, its minimal tension set and compression set result in very little deformation from its original configuration. It does not contain plasticizers that are easily extracted by oils, and yet it has excellent processability. In order to completely satisfy the general requirements for an oil seal material it was necessary to have a different product grade than common silicone rubber. For this reason, the "Oil-Seal Use" grade was developed. Careful consideration was also required in manufacturing processes and quality control. With the improved performance of automobiles, the properties of silicone

can still be utilized for further improvement.
Thus, silicone oil seals are one factor contributing
to the quality improvement of Japanese-made automobiles.
     In the field of color television sets, the RTV
(Room Temperature Vulcanizing) silicone potting agent
of the addition reaction type is used as an insulating
material in high voltage circuits and flyback transformers. With the television receiver having changed
from the monochrome system to the polychrome system
and from the vacuum tube to the transistor, insulation
in high voltage circuits has been the subject of much
research. As shown in Fig. 1, silicone RTV rubber
has two types; the one-component type and the two-
component type. The crosslinking methods include the
condensation reaction and the addition reaction.
At that time, the two-component type based upon the condensation reaction was used as the potting agent.
Two-component RTV rubber based upon the condensation
reaction has one shortcoming: alcohol is generated
during the curing process which lowers insulation
performance. This meant that its performance was
unsatisfactory for insulation in high voltage circuits. The essential requirements for insulating
materials in high voltage circuits include stable
resistivity, flame resistance and heat dissipation.
Other important requirements for practical use are
the viscosity-lowering and fast-curing characteristics.

CROSSLINKING REACTION OF
LIQUID SILICONE RUBBER

One Component :

$$\equiv SiOH + XSi\equiv \longrightarrow \equiv SiOSi\equiv + HX$$

  HX : Acetic Acid, Oxime, Alcohol, Amide,
  Acetone, Amine

■ Two Component :

  ● Condensation Method

$$\equiv SiOH + ROSi\equiv \longrightarrow \equiv SiOSi\equiv + ROH$$

   ROH : Water, Alcohol, Hydrogen

  ● Addition Method

$$\equiv SiCH=CH_2 + HSi\equiv \longrightarrow \equiv SiCH_2CH_2Si\equiv$$

Figure 1

To meet these requirements, studies were conducted on the addition reaction as a new curing process, resulting in the successful development of a silicone potting agent that can fulfill all the requirements. The newly developed potting agent was a major factor in the development of solid-state color television sets, and led to the great international reputation of Japanese color television sets.

The 1970s

In this period, the Japanese economy sustained the heavy blows of the oil shocks, but for all that a steady growth was observed mainly in new industries such as ICs, computers, VTRs, precision machinery, office equipment and so on. Despite temporary stagnation, the production of silicone rapidly recovered, and since then, has enjoyed steady growth. Silicone is quite different on this point from other widely used plastic materials such as polyvinyl chloride and polyethylene. This was, in part, attributed to the established superiority in cost-performance analysis of silicone in comparison with other petrochemicals, but it was also, in large part, the result of an improved capability to meet the high-technology-oriented trends of the Japanese industries and to completely fulfill their needs. Silicone applications representative of the 70´s include rubber contacts for electronic calculators, silicone rubber rolls for Plain Paper Copiers and lubricants for the magnetic tape of VTR´s and audio-equipment.

Earlier calculators utilized spring-loaded mechanical contacts in their keyboard switches, but with the emerging need for a reduced size and cost and for a higher production volume, silicone rubber contacts were developed. The conducting parts are made of conductive silicone rubber mixed with carbon for contacts, and the insulating parts are made of silicone rubber mixed with silica. The required characteristics for rubber contacts included a superior switching function as well as durability. In terms of the physical property requirements of rubber, high resilience, little creeping and small variation in modulus should be included. Silicone rubber contacts are utilized not only in calculators, but also in the channel selectors of television sets, push phones and personal computers and electronic musical instruments. Conductive silicone rubber is also used as the liquid crystal display connector in digital wrist watches, and thus, has contributed to the widespread availability of those watches. Furthermore, it has a wide variety of other applications ranging from various types of sensors to gaskets for computer rooms as an EMI material.

In Plain Paper Copiers (PPC), silicone rubber is utilized in the fuser roll and backup roll. In this application, a choice is made between two types of silicone rubber; two-component RTV rubber or a heat curable rubber compound, depending upon the required characteristics. Japanese copiers have significantly improved in performance. This is partly due to the improved feature of silicone which can meet the rigorous requirements for higher resolution of copy images and prolonged service life, etc. Improvements were made not only in silicone materials but in processing techniques, and the service life of silicone rubber rolls expressed by the number of copies has been extended from a previous 2,000 copies to the current 300,000 copies.

Magnetic recording tape utilized in VTR and audio equipment are required to have sufficient durability as well as a small friction coefficient and better lubricity because they come in a physical contact with tape guides, magnetic heads, and the like during use. A variety of lubricants were used to lubricate magnetic tape, including conventional silicone fluids. However, with the improved performance of magnetic tape, there was an increasing need for development of a much more lubricative silicone fluid. In this way, the optimum molecular design for magnetic tape lubricants was researched, and a large number of special silicone fluids were synthesized and evaluated. This resulted in the successful development of highly durable VTR tape capable of providing high quality images and high fidelity audio tapes.

The 1980s

Regarding the outlook for the silicone industry in terms of application, its market is estimated to be made up as follows; 25% for electric and electronics industries; 15% for construction; 10% for the automobile industry; 10% for office equipment; 10% for medical and food processing and 30% for paper, textile, chemical and others. The industrial sectors that have experienced the most remarkable growth are the electric and electronics sector and the office equipment sector.

The electronics industry has rapidly expanded its applications, leading the on-going changeover to electronics systems being implemented by various industrial segments.

Applications of silicone in electronics include coating for surface protection of semiconductor devices and a plastic package of such discrete devices as transistors, diodes etc. In the lithography process of IC fabrication, hexamethyldisilazane is also applied as an adhesive for photoresists. Silicone is also used in the potting of hybrid IC´s and in the conformal coating of printed circuits. Silicone gel is utilized as a countermeasure to soft errors caused by alpha rays. Silicone is also applied as a coating material for optical fibers utilized in opto-electronics communications, taking advantage of its high transparency, flexibility and purity. In the new electronics equipment, special silane is utilized in orientation of liquid crystal for the displays of miniature pocket TV sets, and silicone gel for video projectors. In the field of medical electronics equipment, silicone rubber is used in the probes of supersonic diagonostic devices. Obviously there is a wide range of silicone applications in the electric and electronics industries.

Silicone is a high performance material, and even though it is now utilized in a wide variety of fields, it is expected to have new applications in the same fields when combined with new technologies. To name a few examples, such applications include the FIPG(Formed-In-Place Gasket) system and brake oil and hard surfacing treatment of plastics in the automobile industry. In the cosmetics, textile and medical industries, there are also needs for new types of products with higher value added.

TRENDS IN THE DEVELOPMENT OF SILICONE TECHNOLOGY

Recent trends in the field of silicone are reviewed in the following from a technical viewpoint. This includes the polymer modification, crosslinking reaction, blending technology and system technology.

Polymer modification

For the purpose of improving silicone characteristics and providing additional functions, a great deal of research has been conducted about the

Possibility of copolymerizing a silicone polymer with an organic polymer. In addition to siloxane polymers, there are a large number of other silicon containing polymers. Research in this area is now being actively conducted, and excellent results are expected in the future. Of the many possibilities, hard contact lenses were realized by copolymerization of silicone and PMMA and covulcanizing rubber was realized by blending silicone and EPDM. These two cases require some explanation.

When acryl-functional siloxane and methylmeth-acrylate are subjected to radical copolymerization, a copolymer is synthesized, with a siloxane chain incorporated in its side chains.

$$(CH_3)_3SiOSiCH_2OCC=CH_2 \quad\quad [1]$$

with methyl groups CH_3 on the silicon and CH_3 on the carbon, and CH_3 and O below.

$$((CH_3)_3SiO)_3Si(CH_2)_3OCH_2CHCH_2\ CH_3 \quad\quad [2]$$

with HO, $O-C-C=CH_2$, and O below.

Due to the incorporation of siloxane, this polymer has about 100 times more oxygen permeability than PMMA, and therefore found a practical use as hard contact lenses. This application exemplifies the excellent gas permeability of the siloxane polymer.

Covulcanizing rubber (SEP rubber) composed of a blend of silicone and EPT or EPDM has characteristics of both silicone rubber and EPDM, but is superior to EPDM in heat resistance, weather resistance and compression set ,and superior to silicone rubber in mechanical strength, hot water resistance and steam resistance. Its special advantage is that flame resistance is obtained without the addition of a flame retardant. SEP rubber can be vulcanized either through a sulfur vulcanization process or a peroxide vulcanization process, allowing the application of various fabrication methods. It is being effectively used as heat-resistant rubber in fields which are between silicone and EPDM in temperature charac-teristics. [3]

Crosslinking reaction

Silicone crosslinking reactions include radical reaction, condensation reaction, addition reaction, etc. One example of condensation reaction, acetone type one-component RTV rubber, is described here. (See Fig. 1)

HIGH TEMPERATURE STABILITY OF R T V

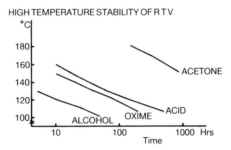

After the uncured RTV compositions are heated at high temperatures, the longest heating time that they can be ordinarily cured at are measured.

Figure 2

Acetone type one-part RTV rubber utilizes organo-propenoxysilane as a crosslinking agent and generates acetone in the curing process. [4] Acetone type RTV rubber has several excellent characteristics suited to electronics use, including fast curing, odorlessness, good adhesiveness, non-corrosiveness, etc. The most important characteristic is reversion resistance. (See Fig. 2)
In some cases, exposure of one-component RTV rubber in an uncured state to a high temperature prevents the rubber from curing or deteriorates its physical properties significantly. This is because, due to the effects of acids, bases and/or catalysts existing in the compositions, siloxane chains are cleved. Acetone type RTV rubber is far superior to other types of one-component RTV rubber on this point. As a result of this feature, the FIPG has advanced. FIPG is a system in which RTV rubber is coated for use as a gasket in automobile production lines to replace conventional prefabricated gaskets made of cork, rubber and/or paper. It has brought about improved reliability of sealing and a substantial reduction in production costs. In automobile fabrication lines, engines are normally tested, and in some cases, under the influence of such tests, RTV

rubber was exposed to a high temperature prior to its being completely cured. Acetone type RTV rubber can be effectively used even in these cases, allowing changeover of the FIPG process to an on-line system.

Liquid silicone rubber, utilizing the addition reaction as a crosslinking reaction is being considered for practical use in combination with curing processes utilizing ultra violet rays or an electron beam. Fig. 3 illustrates proposed UV curing reactions. The curing processes utilizing UV rays or an electron beam are most definitely characterized by fast curing of only several seconds. Expanded use is expected for this material as the coating agent of optical fibers [5], in electronic components and for release paper.

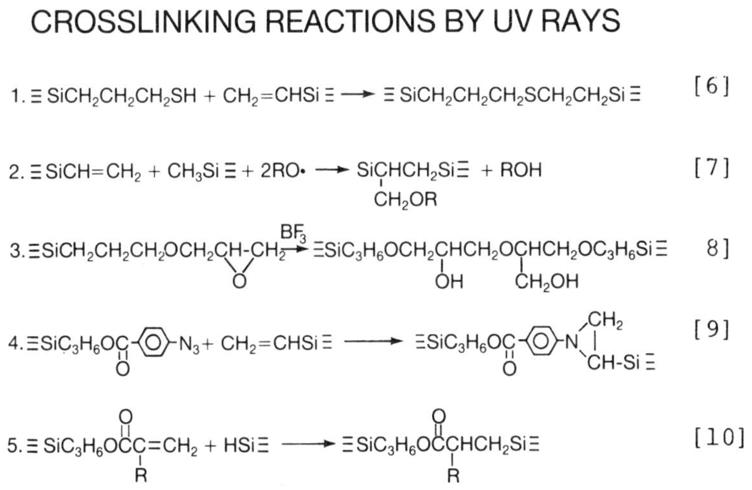

CROSSLINKING REACTIONS BY UV RAYS

1. $\equiv SiCH_2CH_2CH_2SH + CH_2{=}CHSi \equiv \longrightarrow \equiv SiCH_2CH_2CH_2SCH_2CH_2Si \equiv$ [6]

2. $\equiv SiCH{=}CH_2 + CH_3Si \equiv + 2RO\cdot \longrightarrow Si\overset{|}{C}HCH_2Si\equiv + ROH$ [7]
   $\qquad\qquad\qquad\qquad\qquad\qquad\underset{CH_2OR}{}$

3. $\equiv SiCH_2CH_2CH_2OCH_2\overset{|}{C}H\text{-}CH_2 \overset{BF_3}{\longrightarrow} \equiv SiC_3H_6OCH_2\overset{|}{C}HCH_2O\overset{|}{C}HCH_2OC_3H_6Si\equiv$ 8]
   $\qquad\qquad\qquad\qquad\qquad\underset{O}{\diagdown\diagup}\qquad\qquad\qquad\underset{OH}{}\quad\underset{CH_2OH}{}$

4. $\equiv SiC_3H_6O\overset{\underset{||}{O}}{C}\text{-}\!\!\bigcirc\!\!\text{-}N_3 + CH_2{=}CHSi \equiv \longrightarrow \equiv SiC_3H_6O\overset{\underset{||}{O}}{C}\text{-}\!\!\bigcirc\!\!\text{-}N\overset{\diagup CH_2}{\underset{\diagdown CH\text{-}Si\equiv}{\big|}}$ [9]

5. $\equiv SiC_3H_6O\overset{\overset{O}{||}}{C}\overset{|}{C}{=}CH_2 + HSi\equiv \longrightarrow \equiv SiC_3H_6O\overset{\overset{O}{||}}{C}\overset{|}{C}HCH_2Si\equiv$ [10]
   $\qquad\qquad\quad\underset{R}{}\qquad\qquad\qquad\qquad\qquad\underset{R}{}$

Figure 3

Blending technology

Silicone is usually used in a compound form. Therefore, blending technology plays an important role in improving the properties of silicone. To name a few examples, improved mechanical strength and improved flame resistance are the results of efforts to improve compounding techniques. Furthermore, it is possible to give additional functions to silicone through advanced compound techniques, and as mentioned earlier, conductive silicone rubber filled with a conductive filler has found extensive uses.

Research is underway to improve thermal conductivity and to give an antithrombogenic feature to silicone. Active research will continue on blending technology to meet the advanced needs of the marketplace and to open up new markets in the future, too.

System technology

In silicone technology, importance is given not only to the development of materials but to the development of system technology in which the developed materials are used. In parallel with FIPG´s, LIM (Liquid Injection Molding) has spread at a remarkable pace. LIM is an injection system utilizing liquid silicone rubber. LIM is a labor- and energy-saving system, and it has the possibility of a fully automated system through introduction of robots. It mainly utilizes RTV rubber of an addition reaction type as the molding material.

FUTURE PROSPECTS FOR SILICONE

The future growth of silicone probably depends on finding needs in the frontier technologies and developing silicone that will meet those needs. In the fields of electronics, energy and biotechnology, research and development is under way aiming at future technological innovation. A few examples of such research and development should be described.

SILICONE PHOTORESISTS

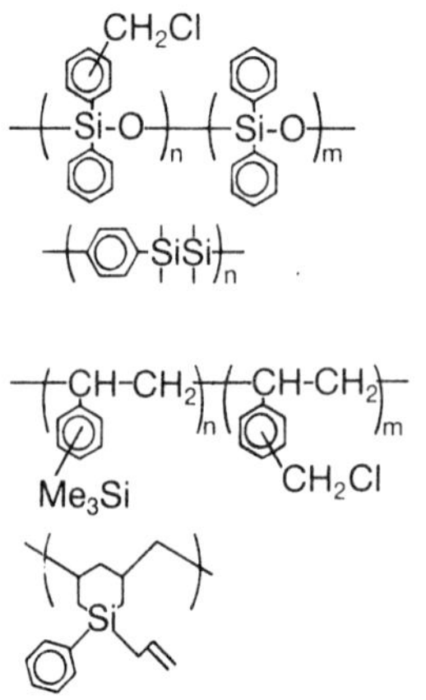

Chloromethylated polydiphenylsiloxane   [11]

Poly(p-disilanylenephenylene) [12]

Poly(trimethylsilylstyrene-chloromethylstyrene)   [13]

Poly(triallylphenylsilane)   [14]

Figure 4

In the field of electronics, silicone has been
used as an electric insulating material, but current-
ly, electronics materials with additional functions
are being developed. These electronics materials
include silicone photoresists   for the manufacturing
of VLSI´s, liquid crystal with a siloxane chain incorporated [15] and conductive polymers [16]. Fig. 4
illustrates a silicone photoresist. Silicone is
resistant to oxygen plasma and can be subjected to
fine fabrication on the order of submicron by means
of Reactive Ion Etching.

In the field of energy, monosilane and disilane
have been developed for use in amorphous silicon
solar cells. Amorphous silicon is also expected to
be applied to photosensitive drums for PPC copiers,
sensors, and image pickup tubes.

In the fields of biotechnology and life science,
development of biocompatible materials and silicone
applications to drug delivery systems and to oxygen
permeable membrances are in progress.
Octadecylsilane is used for surface treatment of
pharmaceutical glassware and appratus, and silane
derivatives are also applicable to immobilized enzyme.
These fields are still unexplored by the silicone
industry, but they have infinite potential.
Silicon compounds are utilized for development of
advanced materials. In the field of ceramics, it is
attracting attention as the raw materials of silicon
nitride and silicon carbide as well as binders, and
silicon carbide fibers have been successfully developed by means of thermal decomposition of organosilicon compounds. Application of silicon reagents
to organic synthesis is widespread, and silicon
reagents are also being applied to polymer synthesis.
Group Transfer Polymerization is one of the examples.
Expectation is being entertained of development of
new reaction processes and materials, making use of
the properties of silicon which differ from the
element of carbon.

It is our great pleasure that silicone has made
a contribution toward development of industry and the
improvement of human life, and we hope a continued
contribution can be made through the advancement of
research and development. To this end, efforts will
be made to improve the properties of silicone to meet
the emerging needs of the frontier technologies.
Organosilicon chemists are requested to keep pace
through expanded dialogue with industrial and academic
circles.

REFERENCES

[1] N. G. Gaylord, US Patent 3,808,178(1974)
[2] K. Tanaka et al., US Patent 4,139,513(1979)
[3] K. Itoh et al., US Patent 4,150,010(1979)
[4] T. Takago et al., US Patent 3,819,563(1974)
[5] T. Kimura and S. Sakaguchi, Electronics Letters, 20 (8), 315(1984)
[6] G. N. Bokerman et al., US Patent 4,052,529(1977)
[7] H. Okinoshima et al., US Patent 4,364,809(1982)
[8] J. V. Crivello and S. H. Schroeter, US Patent 4,026,705(1977)
[9] T. Tsunoda, T. Yamaoka et al., Japan Patent Kokai 54/70104(1979)
[10] Raymond Pigeon, France Patent 2,507,608(1982)
[11] M. Morita et al., Japanese J. Applied Phys., 22 (10), L569(1983)
M. Morita et al., J. Electrochem. Soc., 131, 653(1984)
[12] M. Ishikawa et al., in Proc. 31th Spring Meeting Japan Soc. Appl. Phys., 1a-W-1, p272(1984)
[13] N. Suzuki et al., in Proc. 44th Autumn Meeting Japan Soc. Appl. Phys., 26a-U-7, p258(1983)
[14] N. Suzuki et al., in Proc. 31th Spring Meeting Japan Soc. Appl. Phys., 1a-W-6, p274(1984)
[15] H. Finkelmann et al., Makromol. Chem., Rapid Commun., 5, 287(1984)
[16] T. J. Marks et al., J. Am. Chem. Soc., 105 1539(1983)

Index